PRACTICE · ASSESS

180 Days of GEOGRAPHY
for Second Grade

Author
Melissa Callaghan, NBCT

Series Consultant

Nicholas Baker, Ed.D.
Supervisor of Curriculum and Instruction
Colonial School District, DE

Publishing Credits

Corinne Burton, M.A.Ed., *Publisher*
Conni Medina, M.A.Ed., *Managing Editor*
Emily R. Smith, M.A.Ed., *Content Director*
Veronique Bos, *Creative Director*
Shaun N. Bernadou, *Art Director*
Lynette Ordoñez, *Editor*
Kevin Pham, *Graphic Designer*
Stephanie Bernard, *Associate Editor*

Image Credits

p.58 Library of Congress [LC-USZ62-24396]; p.93 (top) Library of Congress [LC-USZ62-137464]; p.103 (left) Library of Congress [LC-DIG-stereo-1s06734]; p.173 Library of Congress [LC-USZ62-66669]; all other images from iStock and/or Shutterstock.

Standards

© 2012 National Council for Geographic Education
© 2014 Mid-continent Research for Education and Learning (McREL)

> For information on how this resource meets national and other state standards, see pages 10–14. You may also review this information by visiting our website at www.teachercreatedmaterials.com/administrators/correlations/ and following the on-screen directions.

Shell Education

A division of Teacher Created Materials
5301 Oceanus Drive
Huntington Beach, CA 92649-1030
www.tcmpub.com/shell-education

ISBN 978-1-4258-3303-9

©2018 Shell Educational Publishing, Inc.

The classroom teacher may reproduce copies of materials in this book for classroom use only. The reproduction of any part for an entire school or school system is strictly prohibited. No part of this publication may be transmitted, stored, or recorded in any form without written permission from the publisher.

28623—*180 Days of Geography*

© *Shell Education*

TABLE OF CONTENTS

Introduction 3

How to Use This Book 4

Standards Correlations 10

Daily Practice Pages 15

Answer Key 195

Rubrics 202

Analysis Pages 205

Digital Resources 208

INTRODUCTION

With today's geographic technology, the world seems smaller than ever. Satellites can accurately measure the distance between any two points on the planet and give detailed instructions about how to get there in real time. This may lead some people to wonder why we still study geography.

While technology is helpful, it isn't always accurate. We may need to find detours around construction, use a trail map, outsmart our technology, and even be the creators of the next navigational technology.

But geography is also the study of cultures and how people interact with the physical world. People change the environment, and the environment affects how people live. People divide the land for a variety of reasons. Yet no matter how it is divided or why, people are at the heart of these decisions. To be responsible and civically engaged, students must learn to think in geographical terms.

The Need for Practice

To be successful in geography, students must understand how the physical world affects humanity. They must not only master map skills but also learn how to look at the world through a geographical lens. Through repeated practice, students will learn how a variety of factors affect the world in which they live.

Understanding Assessment

In addition to providing opportunities for frequent practice, teachers must be able to assess students' geographical understandings. This allows teachers to adequately address students' misconceptions, build on their current understandings, and challenge them appropriately. Assessment is a long-term process that involves careful analysis of student responses from a discussion, project, practice sheet, or test. The data gathered from assessments should be used to inform instruction: slow down, speed up, or reteach. This type of assessment is called *formative assessment*.

HOW TO USE THIS BOOK

Weekly Structure

The first two weeks of the book focus on map skills. By introducing these skills early in the year, students will have a strong foundation on which to build throughout the year. Each of the remaining 34 weeks will follow a regular weekly structure.

Each week, students will study a grade-level geography topic and a location in North America. Locations may be a town, a state, a region, or the whole continent.

Days 1 and 2 of each week focus on map skills. Days 3 and 4 allow students to apply information and data to what they have learned. Day 5 helps students connect what they have learned to themselves.

 Day 1—Reading Maps: Students will study a grade-appropriate map and answer questions about it.

 Day 2—Creating Maps: Students will create maps or add to an existing map.

 Day 3—Read About It: Students will read a text related to the topic or location for the week and answer text-dependent or photo-dependent questions about it.

 Day 4—Think About It: Students will analyze a chart, diagram, or other graphic related to the topic or location for the week and answer questions about it.

 Day 5—Geography and Me: Students will do an activity to connect what they learned to themselves.

Five Themes of Geography

Good geography teaching encompasses all five themes of geography: location, place, human-environment interaction, movement, and region. Location refers to the absolute and relative locations of a specific point or place. The place theme refers to the physical and human characteristics of a place. Human-environment interaction describes how humans affect their surroundings and how the environment affects the people who live there. Movement describes how and why people, goods, and ideas move between different places. The region theme examines how places are grouped into different regions. Regions can be divided based on a variety of factors, including physical characteristics, cultures, weather, and political factors.

HOW TO USE THIS BOOK (cont.)

Weekly Themes

The following chart shows the topics, locations, and themes of geography that are covered during each week of instruction.

Week	Topic	Location	Theme(s)
1	—Map Skills Only—		Location
2			Location
3	Types of Maps	Community	Location, Place
4	Transportation	North America	Movement
5	Water	North America	Human-Environment Interaction
6	International Borders	North America	Location, Place
7	Rural, Urban, and Suburban Communities	United States	Place
8	Landform	Great Basin	Location, Place, Region
9	German Immigration	Midwest	Movement
10	Landform	Atlantic and Gulf Coastal Plain	Region
11	Population	United States	Location, Movement
12	Landform	Mississippi River	Movement, Human-Environment Interaction
13	Landform	Great Lakes	Place, Human-Environment Interaction
14	Communities	Town	Location, Place
15	Water	North America	Human-Environment Interaction
16	Changing Communities	Las Vegas	Place, Human-Environment Interaction
17	Transportation	Panama Canal	Location, Movement
18	Electricity	New York	Place, Movement

HOW TO USE THIS BOOK (cont.)

Week	Topic	Location	Theme(s)
19	Landform	Gulf of Mexico	Human-Environment Interaction, Region
20	European Settlement	United States	Human-Environment Interaction, Movement
21	Regions	United States	Movement, Region
22	Landform	Continental Divide	Human-Environment Interaction, Movement
23	Culture	Alaska	Place, Human-Environment Interaction
24	Transportation	Erie Canal	Movement
25	Growing Corn	Midwest	Human-Environment Interaction, Region
26	Renewable and Non-renewable Resources	United States	Human-Environment Interaction
27	Culture	Aztec Empire	Place, Movement
28	Landform	Bering Strait	Location, Place
29	Culture	Canada	Place
30	Estuary	Chesapeake Bay	Human-Environment Interaction
31	Landform	Colorado River	Human-Environment Interaction
32	American Indians	Great Plains	Human-Environment Interaction, Region
33	Transcontinental Railroad	Western United States	Movement
34	Population	United States	Movement, Region
35	Ecosystem	Yellowstone National Park	Location, Human-Environment Interaction
36	Climate	United States	Human-Environment Interaction, Region

HOW TO USE THIS BOOK (cont.)

Using the Practice Pages

The activity pages provide practice and assessment opportunities for each day of the school year. Teachers may wish to prepare packets of weekly practice pages for the classroom or for homework.

As outlined on page 4, each week examines one location and one geography topic.

The first two days focus on map skills. On Day 1, students will study a map and answer questions about it. On Day 2, they will add to or create a map.

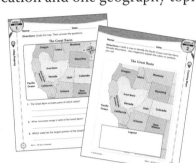

Days 3 and 4 allow students to apply information and data from texts, charts, graphs, and other sources to the location being studied.

On Day 5, students will apply what they learned to themselves.

Using the Resources

Rubrics for the types of days (map skills, applying information and data, and making connections) can be found on pages 202–204 and in the Digital Resources. Use the rubrics to assess students' work. Be sure to share these rubrics with students often so that they know what is expected of them.

© Shell Education

28623—180 Days of Geography

HOW TO USE THIS BOOK (cont.)

Diagnostic Assessment

Teachers can use the practice pages as diagnostic assessments. The data analysis tools included with the book enable teachers or parents to quickly score students' work and monitor their progress. Teachers and parents can quickly see which skills students may need to target further to develop proficiency.

Students will learn map skills, how to apply information and data, and how to relate what they learned to themselves. You can assess students' learning in each area using the rubrics on pages 202–204. Then, record their scores on the Practice Page Item Analysis sheets on pages 205–207. These charts are also provided in the Digital Resources as PDFs, Microsoft Word® files, and Microsoft Excel® files (see page 208 for more information). Teachers can input data into the electronic files directly on the computer, or they can print the pages.

To Complete the Practice Page Item Analyses:

- Write or type students' names in the far-left column. Depending on the number of students, more than one copy of the forms may be needed.
    - The skills are indicated across the tops of the pages.
    - The weeks in which students should be assessed are indicated in the first rows of the charts. Students should be assessed at the ends of those weeks.
- Review students' work for the days indicated in the chart. For example, if using the Making Connections Analysis sheet for the first time, review students' work from Day 5 for all five weeks.
- Add the scores for each student. Place that sum in the far right column. Record the class average in the last row. Use these scores as benchmarks to determine how students are performing.

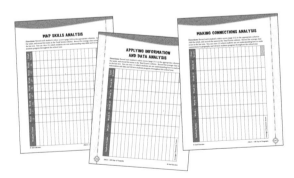

Digital Resources

The Digital Resources contain digital copies of the rubrics, item analysis sheets, and standards charts. See page 208 for more information.

HOW TO USE THIS BOOK (cont.)

Using the Results to Differentiate Instruction

Once results are gathered and analyzed, teachers can use them to inform the way they differentiate instruction. The data can help determine which geography skills are the most difficult for students and which students need additional instructional support and continued practice.

Whole-Class Support

The results of the diagnostic analysis may show that the entire class is struggling with certain geography skills. If these concepts have been taught in the past, this indicates that further instruction or reteaching is necessary. If these concepts have not been taught in the past, this data is a great preassessment and may demonstrate that students do not have a working knowledge of the concepts. Thus, careful planning for the length of the unit(s) or lesson(s) must be considered, and additional front-loading may be required.

Small-Group or Individual Support

The results of the diagnostic analysis may show that an individual student or a small group of students is struggling with certain geography skills. If these concepts have been taught in the past, this indicates that further instruction or reteaching is necessary. Consider pulling these students aside to instruct them further on the concepts while others are working independently. Students may also benefit from extra practice using games or computer-based resources.

Teachers can also use the results to help identify proficient individual students or groups of students who are ready for enrichment or above-grade-level instruction. These students may benefit from independent learning contracts or more challenging activities.

STANDARDS CORRELATIONS

Shell Education is committed to producing educational materials that are research and standards based. In this effort, we have correlated all our products to the academic standards of all 50 states, the District of Columbia, the Department of Defense Dependents Schools, and all Canadian provinces.

How to Find Standards Correlations

To print a customized correlation report of this product for your state, visit our website at **www.teachercreatedmaterials.com/administrators/correlations** and follow the on-screen directions. If you require assistance in printing correlation reports, please contact our Customer Service Department at 1-877-777-3450.

Purpose and Intent of Standards

The Every Student Succeeds Act (ESSA) mandates that all states adopt challenging academic standards that help students meet the goal of college and career readiness. While many states already adopted academic standards prior to ESSA, the act continues to hold states accountable for detailed and comprehensive standards. Standards are designed to focus instruction and guide adoption of curricula. Standards are statements that describe the criteria necessary for students to meet specific academic goals. They define the knowledge, skills, and content students should acquire at each level. Standards are also used to develop standardized tests to evaluate students' academic progress. Teachers are required to demonstrate how their lessons meet state standards. State standards are used in the development of our products, so educators can be assured they meet the academic requirements of each state.

The activities in this book are aligned to the National Geography Standards and the McREL standards. The chart on pages 11–12 lists the National Geography Standards used throughout this book. The chart on pages 13–14 correlates the specific McREL and National Geography Standards to each week. The standards charts are also in the Digital Resources (standards.pdf).

C3 Framework

This book also correlates to the College, Career, and Civic Life (C3) Framework published by the National Council for the Social Studies. By completing the activities in this book, students will learn to answer and develop strong questions (Dimension 1), critically think like a geographer (Dimension 2), and effectively choose and use geography resources (Dimension 3). Many activities also encourage students to take informed action within their communities (Dimension 4).

STANDARDS CORRELATIONS *(cont.)*

180 Days of Geography is designed to give students daily practice in geography through engaging activities. Students will learn map skills, how to apply information and data to their understandings of various locations and cultures, and how to apply what they learned to themselves.

Easy to Use and Standards Based

There are 18 National Geography Standards, which fall under six essential elements. Specific expectations are given for fourth grade, eighth grade, and twelfth grade. For this book, fourth grade expectations were used with the understanding that full mastery is not expected until that grade level.

Essential Elements	National Geography Standards
The World in Spatial Terms	**Standard 1:** How to use maps and other geographic representations, geospatial technologies, and spatial thinking to understand and communicate information
	Standard 2: How to use mental maps to organize information about people, places, and environments in a spatial context
	Standard 3: How to analyze the spatial organization of people, places, and environments on Earth's surface
Places and Regions	**Standard 4:** The physical and human characteristics of places
	Standard 5: People create regions to interpret Earth's complexity
	Standard 6: How culture and experience influence people's perceptions of places and regions
Physical Systems	**Standard 7:** The physical processes that shape the patterns of Earth's surface
	Standard 8: The characteristics and spatial distribution of ecosystems and biomes on Earth's surface

STANDARDS CORRELATIONS (cont.)

Essential Elements	National Geography Standards
Human Systems	**Standard 9:** The characteristics, distribution, and migration of human populations on Earth's surface
Human Systems	**Standard 10:** The characteristics, distribution, and complexity of Earth's cultural mosaics
Human Systems	**Standard 11:** The patterns and networks of economic interdependence on Earth's surface
Human Systems	**Standard 12:** The process, patterns, and functions of human settlement
Human Systems	**Standard 13:** How the forces of cooperation and conflict among people influence the division and control of Earth's surface
Environment and Society	**Standard 14:** How human actions modify the physical environment
Environment and Society	**Standard 15:** How physical systems affect human systems
Environment and Society	**Standard 16:** The changes that occur in the meaning, use, distribution, and importance of resources
The Uses of Geography	**Standard 17:** How to apply geography to interpret the past
The Uses of Geography	**Standard 18:** How to apply geography to interpret the present and plan for the future

—2012 National Council for Geographic Education

STANDARDS CORRELATIONS (cont.)

Easy to Use and Standards Based (cont.)

This chart lists the specific National Geography Standards and McREL standards that are covered each week.

Wk.	NGS	McREL Standard
1	Standards 1 and 2	Understands the characteristics and uses of maps, globes, and other geographic tools and technologies.
2	Standards 1 and 2	Understands the characteristics and uses of maps, globes, and other geographic tools and technologies.
3	Standards 1 and 2	Knows the location of school, home, neighborhood, community, state, and country.
4	Standard 11	Knows the modes of transportation used to move people, products, and ideas from place to place, their importance, and their advantages and disadvantages.
5	Standard 14	Knows ways that people solve common problems by cooperating.
6	Standards 1 and 13	Knows the location of school, home, neighborhood, community, state, and country.
7	Standards 4 and 12	Knows the similarities and differences in housing and land use in urban and suburban areas.
8	Standards 4 and 5	Knows areas that can be classified as regions according to physical criteria and human criteria.
9	Standard 10	Knows the basic components of culture.
10	Standards 4 and 5	Knows areas that can be classified as regions according to physical criteria and human criteria.
11	Standard 9	Understands why people choose to settle in different places.
12	Standards 7 and 14	Knows the modes of transportation used to move people, products, and ideas from place to place, their importance, and their advantages and disadvantages.
13	Standards 4 and 7	Knows that places can be defined in terms of their predominant human and physical characteristics.
14	Standard 1	Knows the absolute and relative location of a community and places within it.
15	Standard 14	Knows ways in which people depend on the physical environment.
16	Standards 12 and 14	Knows how areas of a community have changed over time.
17	Standard 17	Knows the modes of transportation used to move people, products, and ideas from place to place, their importance, and their advantages and disadvantages.
18	Standard 14	Knows how areas of a community have changed over time.

© Shell Education

STANDARDS CORRELATIONS (cont.)

Wk.	NGS	McREL Standard
19	Standards 5 and 7	Knows areas that can be classified as regions according to physical criteria and human criteria.
20	Standard 17	Knows how areas of a community have changed over time.
21	Standards 5 and 15	Knows areas that can be classified as regions according to physical criteria and human criteria.
22	Standards 4 and 17	Knows that places can be defined in terms of their predominant human and physical characteristics.
23	Standard 15	Knows the basic components of culture.
24	Standard 11	Knows the modes of transportation used to move people, products, and ideas from place to place, their importance, and their advantages and disadvantages.
25	Standard 5	Knows areas that can be classified as regions according to physical criteria and human criteria.
26	Standard 16	Knows the role that resources play in our daily lives.
27	Standards 9 and 10	Knows the basic components of culture.
28	Standards 4 and 7	Knows that places can be defined in terms of their predominant human and physical characteristics.
29	Standard 10	Knows the basic components of culture.
30	Standards 4 and 7	Knows that places can be defined in terms of their predominant human and physical characteristics.
31	Standards 7 and 14	Knows that places can be defined in terms of their predominant human and physical characteristics.
32	Standards 5 and 9	Knows areas that can be classified as regions according to physical criteria and human criteria.
33	Standard 17	Knows the modes of transportation used to move people, products, and ideas from place to place, their importance and their advantages and disadvantages.
34	Standards 3 and 9	Understands why people choose to settle in different places.
35	Standard 8	Knows that places can be defined in terms of their predominant human and physical characteristics.
36	Standard 7	Knows ways in which people depend on the physical environment.

Name: _____ Date: _____

Directions: Maps have titles to tell what they show. A legend tells what the symbols mean. And a compass rose tells directions. Study the map, and answer the questions.

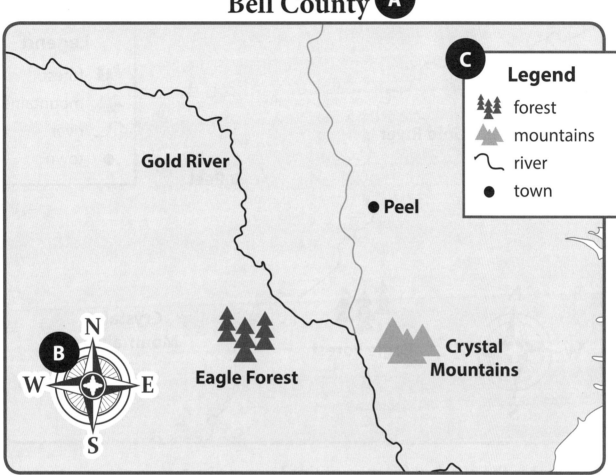

1. What is labeled with letter A?

   _____

2. What is labeled with letter B?

   _____

3. What is labeled with letter C?

   _____

WEEK 1 DAY 2

Name: _____ Date: _____

Directions: Maps use symbols. The legend tells what the symbols mean. Use the legend to answer the questions.

Bell County

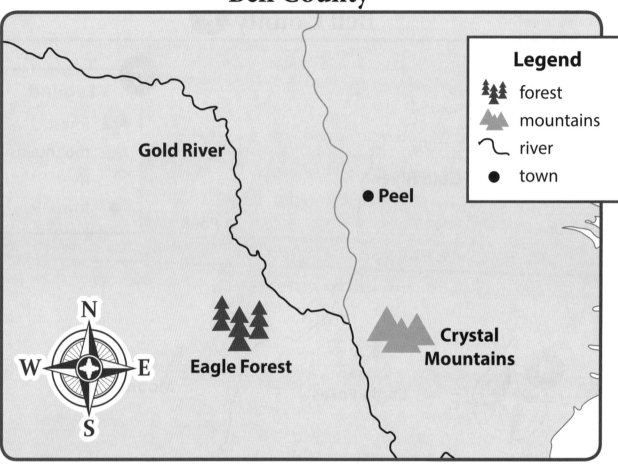

1. What is the symbol for mountains?

   _____

2. What is the symbol for forests?

   _____

3. What is the symbol for a town?

   _____

16   28623—180 Days of Geography   © Shell Education

Name: _____ Date: _____

Directions: A compass rose tells directions on a map. Use the compass rose to answer the questions.

Bell County

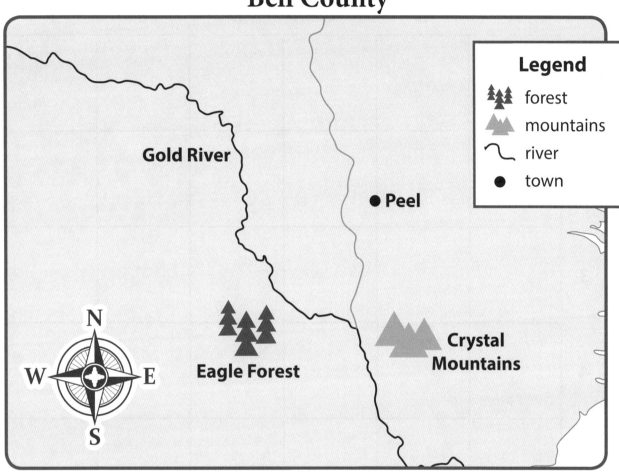

1. What is west of the town of Peel?

   _____

2. If you are in Eagle Forest, what direction do you go to get to the mountains?

   _____

3. What is south of the town of Peel?

   _____

WEEK 1 DAY 4

Name: _____ Date: _____

Directions: Some maps have grids to help people find places. Follow the directions to complete the grid.

Map Skills

	A	B	C	D	E	F
1						
2						
3						
4						
5						
6						

1. Draw a red x in C3.

2. Draw a blue circle in A5.

3. Place an orange triangle in B2.

WEEK 1 DAY 5

Name: _____ **Date:** _____

Directions: A map shows a bird's-eye view of a place. Draw a map of your playground. Include a title and a legend. Use symbols to stand for things in the playground.

Title: _____

Map Skills

WEEK 2 DAY 1

Name: _____ Date: _____

Directions: A globe is a model of Earth. This section of the globe shows North America. Follow the steps to color the globe.

1. Color the United States blue.

2. Color Canada red.

3. Color Mexico green.

Name: _____ **Date:** _____

Directions: Globes show continents and oceans. Use the globe to answer the questions.

1. What ocean is east of the United States?

   _____

2. What ocean is west of the United States?

   _____

3. What ocean is north of Canada?

   _____

WEEK 2 DAY 3

Name: _____ Date: _____

Directions: Use the globe to answer the questions.

1. What country is north of the United States?

   _____

2. What country is south of the United States?

   _____

3. Which of these countries is largest?

   _____

Name: _____ Date: _____

Directions: Some maps show raised places where mountains are. Follow the steps to color the globe.

1. Color the Rocky Mountains orange. They go all the way through the United States and into Canada.

2. Color the Appalachian Mountains brown. They are in the eastern United States.

3. Color the Sierra Madres Mountains green. They are in Mexico.

WEEK 2 DAY 5

Name: _____ Date: _____

Directions: Earth has seven continents. Use the clues to label the continents.

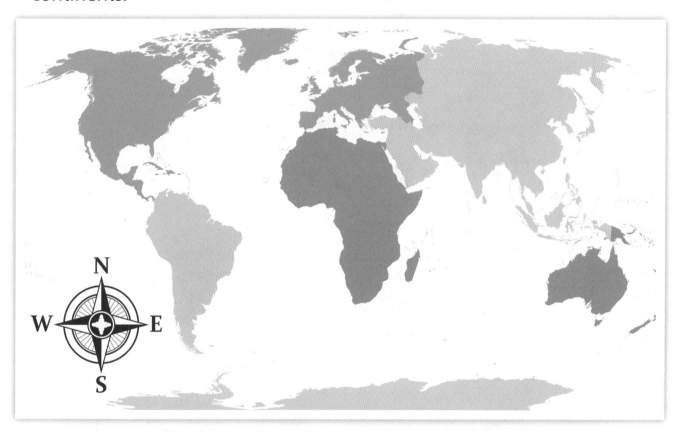

Clues	
Africa is in the center of the map.	Europe is west of Asia.
Antarctica is the farthest south.	North America is on the left side of the map.
Asia is a large continent on the right side of the map.	South America is south of North America.
Australia is the smallest continent.	

Challenge: Place a star on the continent where you live.

Name: _____ **Date:** _____

Directions: Study the map. Then, answer the questions.

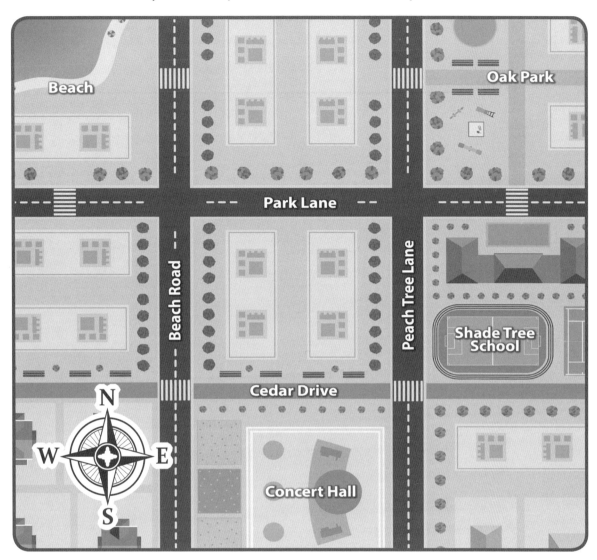

1. What direction is the park from the school?

   _____

2. What directions is the beach from the park?

   _____

3. What direction is the concert hall from the school?

   _____

WEEK 3 DAY 2

Name: _____ **Date:** _____

Directions: Draw a map of your neighborhood. Remember, a map is a bird's-eye view. Include a title and a legend.

Creating Maps

Title: _____

Name: _____ Date: _____

Directions: Read the text. Then, answer the questions.

Maps, Maps, Maps!

You are in a community. A community is often in a state or province. A state or province is in a country. A country is on a continent. A continent is on a planet. You are in a lot of places, all at the same time. That's a lot to keep track of! There are maps for all kinds of places. They are used for different reasons.

A teacher can make a map of a classroom. It can help him or her arrange the desks. You can look at a map of a school. This will tell you where to go in a fire drill. There are maps of your community. They can tell a bus driver where to go. A map of a state or province can help you plan a trip. A map of your country can show places to visit. You can see what countries are near yours by looking at a map of the continent. Maps of Earth show all the continents. A globe is a special kind of map of Earth. It is round, like Earth, instead of being flat. Globes help people understand the world.

1. When might a map of the community be helpful?

   _____

   _____

2. How are a map and a globe different?

   _____

   _____

   _____

WEEK 3 DAY 4

Name: _____ Date: _____

Directions: Study both images. Then, answer the questions.

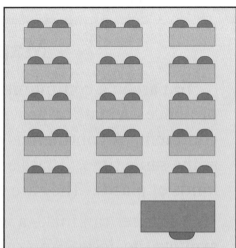

1. How are these images the same?

   _____
   _____
   _____

2. How are these images different?

   _____
   _____
   _____

3. Which image gives you more information? Why?

   _____
   _____
   _____

WEEK 3 DAY 5

Name: _____ **Date:** _____

Directions: A new student has joined your class. Draw a map of your classroom to help her find her seat. Label the map, and include a legend.

Geography and Me

WEEK 4 DAY 1

Name: _____ Date: _____

Directions: Countries in North America trade what they grow with other countries. This map shows what three countries trade. Use the map to answer the questions.

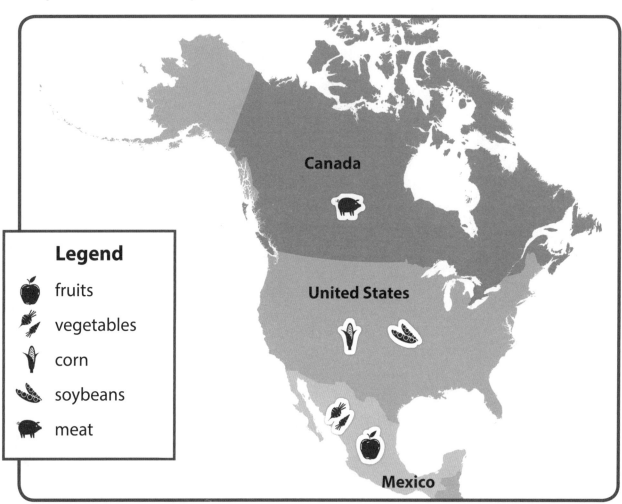

1. What does the United States trade?

   _____

2. What does Mexico trade?

   _____

3. What does Canada trade?

   _____

WEEK 4
DAY 2

Name: _____ Date: _____

Directions: The countries in North America also trade things they make. Study the chart. Add symbols to the map to show what each country trades. Then, add your symbols to the legend.

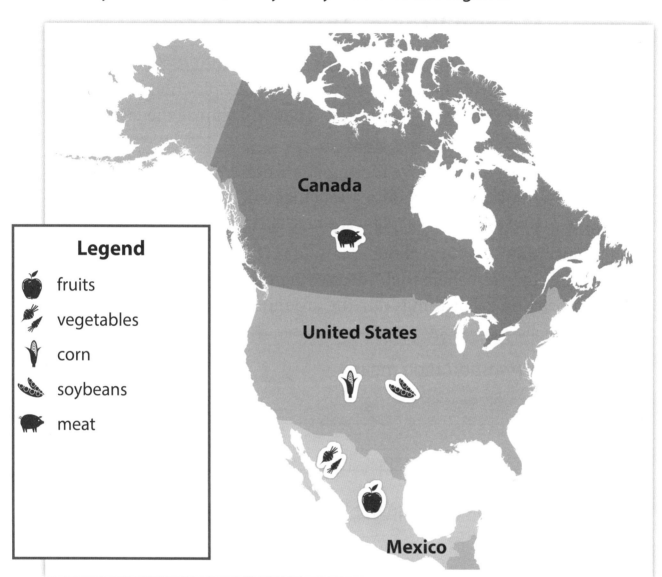

Creating Maps

Country	Products They Trade
Canada	oil
United States	machines
Mexico	cars and trucks

WEEK 4 DAY 3

Name: _____ Date: _____

Directions: Read the text, and study the photo. Then, answer the questions.

Moving People, Things, and Ideas

Transportation is the movement of people or things. People can walk, ride bikes, ride in cars, or fly in planes. People choose how they go. Some people are in a hurry. They will choose the fastest way to go.

Companies need to move their goods around the country and around the world. They choose a way that will save money. They can move things on trucks, planes, ships, and trains.

Transportation is also about ideas. Ideas can spread quickly online. People can share their ideas with others far away. Transportation keeps things moving around our world.

1. What does *transportation* mean?

   _____

   _____

2. How do companies choose the types of transportation for their goods?

   _____

   _____

3. How do people choose the types of transportation they use?

   _____

   _____

WEEK 4 DAY 4

Name: _____ Date: _____

Directions: The chart shows how long it takes to go from Washington, DC, to Los Angeles using different types of transportation. Study the chart, and answer the questions.

Type of Transportation	Time
car	4 to 5 days
airplane	6 hours
train	3 days

1. Do you think the trip by car is long or short? Why?

   _____
   _____
   _____

2. Which type of transportation is the fastest for this trip?

   _____
   _____

3. Which of these types of transportation would you prefer for this trip? Why?

   _____
   _____
   _____
   _____

Think About It

WEEK 4 DAY 5

Name: _____ Date: _____

Directions: Draw and label the types of transportation you have used this week.

WEEK 5 DAY 1

Name: _____ Date: _____

Directions: Follow the steps to color the map.

1. Color the land green.

2. Color the water blue.

3. Is there more land or more water? How do you know?

_____

_____

Reading Maps

WEEK 5 DAY 2

Name: _____ Date: _____

Directions: Use the word bank to label the map.

Word Bank		
Arctic Ocean	Atlantic Ocean	Pacific Ocean

Name: _____ **Date:** _____

Directions: Read the text, and study the photo. Then, answer the questions.

Taking Care of the Water

There is a lot of water in the world. Yet, we can only use a little bit of it. Ocean water is salty. Some water is frozen in the ice caps, too. People need fresh water to drink, grow food, cook, and shower. Many animals need fresh water to drink. Plants need fresh water to grow.

People can protect the water near them. Picking up trash makes the water cleaner. When people recycle, less trash ends up in the water. People can save water by turning off the faucet when they brush their teeth. They can take shorter showers. When people work together, they can protect the water.

1. There is a lot of water on Earth. Why do we need to save water?

   _____

   _____

2. How do picking up trash and recycling help protect water?

   _____

   _____

3. How can people protect the water near them?

   _____

   _____

WEEK 5 DAY 4

Name: _____ Date: _____

Directions: The table shows ways to save water. Study the table, and answer the questions.

Activity	Way to Save	Amount You Could Save
brushing your teeth	turn off the tap while you brush	4 gallons (15 L)
taking a shower	take a shower for less than five minutes	15 gallons (57 L)
taking a bath	only fill the tub halfway	35 gallons (132 L)

1. How much water can you save if you take a short shower?

   _____

2. How much water could you save if you turned off the tap while you brush your teeth?

   _____

3. Which way saves the most water? How do you know?

   _____

   _____

4. What are other ways to save water that are not in the chart?

   _____

   _____

   _____

WEEK 5 DAY 5

Name: _____ **Date:** _____

Directions: One way you can take care of the water near you is by recycling. Glass, plastic, cans, and paper can be recycled. Make a list of the things your family could recycle this week.

Geography and Me

WEEK 6 DAY 1

Name: _____ Date: _____

Directions: Follow the steps to outline the borders of the largest countries in North America. Where countries share a border, you will have two lines right next to each other.

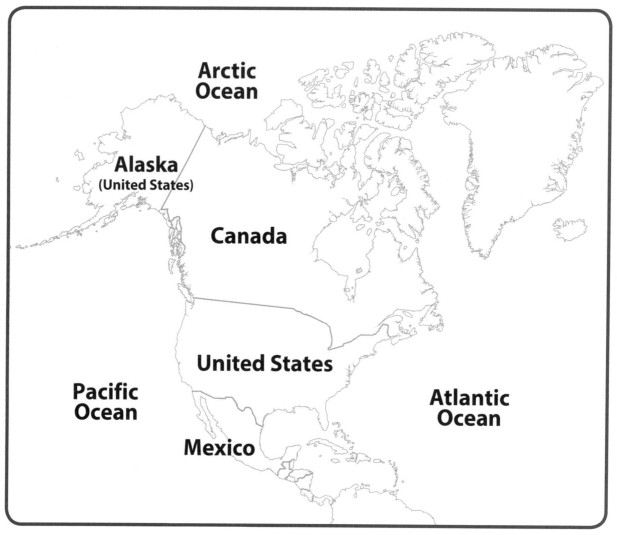

1. Outline Canada in red.

2. Outline the United States in purple.

3. Outline Mexico in green.

Name: _____ **Date:** _____

Directions: On the map, label Canada, the United States, and Mexico.

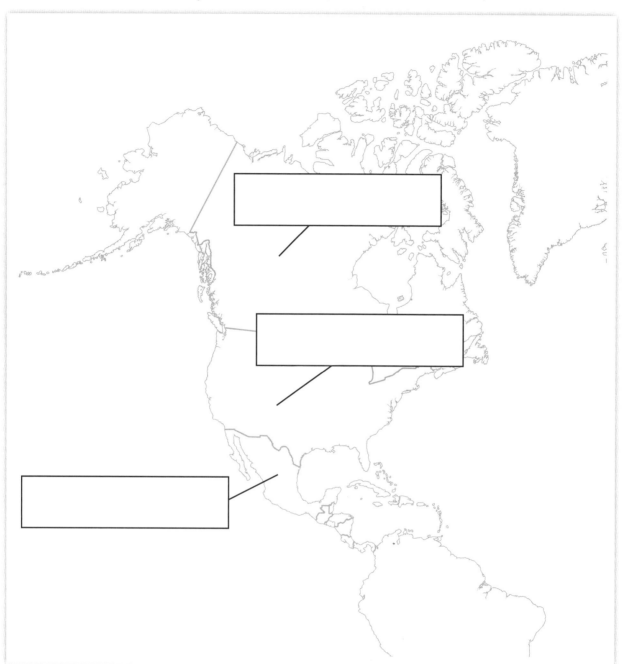

Challenge: Find a place you would like to visit. Place a star on the map above to show where it is. Label the place.

WEEK 6 DAY 3

Name: _____ Date: _____

Directions: Read the text, and study the map. Then, answer the questions.

How Borders Are Made

A border is where a country, state, or city ends. Countries have borders so leaders can make laws for their people. A border can be natural, such as a river. Or it can be a line drawn on a map.

The United States and Canada agreed on a border. They drew a line on a map. The United States and Mexico made a border at the end of a war. They used a river for much of the border. It is called the Rio Grande. People make borders, so borders can change. People may change borders to solve problems.

1. Why do countries have borders?

   _____

   _____

2. What do the United States and Mexico use for a border?

   _____

3. Why would a border change?

   _____

   _____

WEEK 6 DAY 4

Name: _____ **Date:** _____

Directions: Countries can share a border and not speak the same language. Use the chart to answer the questions.

Country	Most Common Language(s)
Canada	English and French
United States	English
Mexico	Spanish

1. Which two countries share a common language?

   _____

   _____

2. What is the most common language spoken in Mexico?

   _____

3. Do you think speaking different languages makes it easier or harder for people in other countries to get along? Why?

   _____

   _____

   _____

   _____

4. What other language would you like to learn? Why?

   _____

   _____

Think About It

WEEK 6
DAY 5

Geography and Me

Name: _____ Date: _____

Directions: Imagine you had to decide how to share the playground with other classes. Draw and write to show how you would make sure the students get along.

_____
_____
_____
_____
_____

Name: _____ Date: _____

Directions: Read the text, and study the photo. Then, answer the questions.

> Urban areas are big cities. Suburban cities have homes and shops spread out over more land. Rural areas have homes and farms spread out over very large areas.

1. What type of community does the cow live in? How do you know?

   _____

   _____

2. What type of community is behind the cow?

   _____

3. What might happen if the city behind the cow expands?

   _____

WEEK 7 DAY 2

Name: _____ **Date:** _____

Directions: This photo shows a place from above. Label the city and the rural area. In the boxes, draw pictures of things you might find in each area.

city	rural area

Name: _____ **Date:** _____

Directions: Read the text. Then, answer the questions.

Urban, Rural, and Suburban Communities

There are different types of communities. In an urban setting, people live very close together. People may walk or ride public transportation to get to work or to shop. They may not need to have a car to get around.

In a rural setting, the homes are far apart. There is a lot of open space. Farmers raise crops and animals here. There are few places to shop. People have to drive to get to places.

There are also suburban settings. Here, the houses are closer together. But they are more spread out than in urban places. People may still have to drive to places. But it does not take as long as in rural places.

1. If you live very close to many people, what kind of community do you live in?

   _____

2. If there is a lot of open space, what kind of community do you live in?

   _____

3. Which two types of communities are the most different? Why?

   _____

   _____

Directions: Study the table to see how some states have changed. Use the table to answer the questions.

State	Number of People in 2000	Number of People in 2010
Nevada	2 million	3 million
Michigan	10 million	9 million
Florida	16 million	19 million

1. Which two states had more people in 2010 than in 2000?

   _____

   _____

2. Which state had fewer people in 2010 than in 2000? How do you know?

   _____

   _____

3. Do you think urban areas in Michigan are getting bigger or smaller? How do you know?

   _____

   _____

   _____

   _____

   _____

WEEK 7 DAY 5

Name: _____ Date: _____

Directions: Draw a picture of where you live. Label it *rural*, *urban*, or *suburban*.

Geography and Me

WEEK 8 DAY 1

Name: _____ Date: _____

Directions: Study the map. Then, answer the questions.

1. The Great Basin includes parts of which states?

   _____

   _____

2. What mountain range is west of the Great Basin?

   _____

3. Which state has the largest portion of the Great Basin?

   _____

WEEK 8 DAY 2

Name: _____ Date: _____

Directions: Create a way to identify the Pacific Ocean and the Sierra Nevada Mountains. Use a legend to explain the colors or symbols you use.

The Great Basin

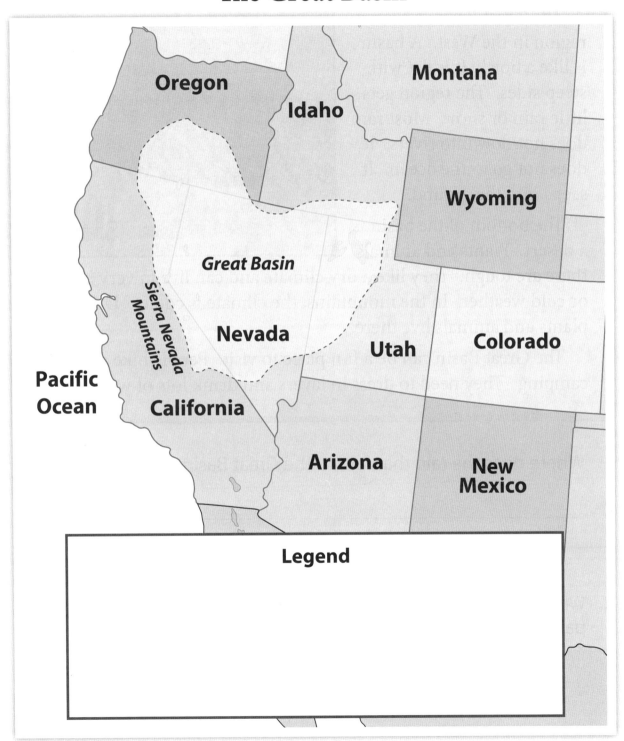

Creating Maps

WEEK 8 DAY 3

Name: _____ Date: _____

Directions: Read the text, and study the photo. Then, answer the questions.

The Great Basin Region

The Great Basin is a large region in the West. A basin is like a bowl. It is flat with steep sides. The region gets little rain or snow. Most rain does not flow into rivers. It does not go to the ocean. It seeps into the ground.

The bottom of the basin is a desert. Plants and animals there are tough. They like a dry climate and can live in very hot or cold weather. In the mountains, the climate is cooler. Different plants and animals live there.

The Great Basin can be a fun place to visit. People hike and go camping. They need to dress in layers and drink lots of water.

1. Where does the rain that falls in the Great Basin region go?

   _____

   _____

2. Why are there different types of plants and animals in different parts of the Great Basin?

   _____

   _____

Name: _____ Date: _____

Directions: *Elevation* means how high the land is above the ocean. The level of the ocean is called *sea level*. The climate changes as the elevation changes. Study the table. Then, answer the questions.

Locations in the Great Basin	Elevation
Badwater Basin	279 ft. (85 m) below sea level
Humboldt Sink	3,894 ft. (1,187 m) above sea level
Mount Whitney	14,505 ft. (4,421 m) above sea level

1. What is the lowest point in the Great Basin region?

   _____

2. What is the highest point in the Great Basin region?

   _____

   _____

3. What is the difference in elevation between Mount Whitney and Humboldt Sink?

   _____

4. Which location do you think is the coldest? Why?

   _____

   _____

Challenge: Find out what the elevation in your area is.

WEEK 8 DAY 5

Name: _____ Date: _____

Directions: Many people go camping in Great Basin National Park. Draw and label the things you might need to go camping there.

Geography and Me

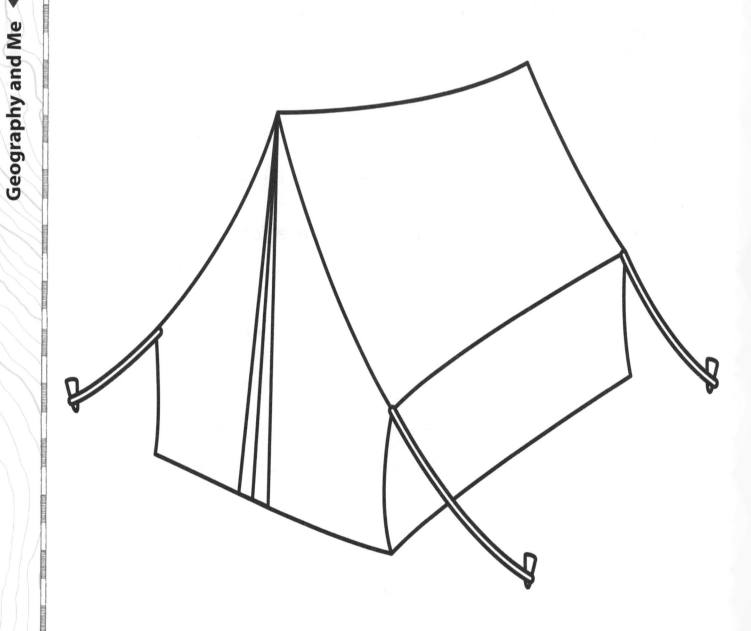

Name: _____ Date: _____

Directions: Many German immigrants settled in the Midwest in the 1800s. Immigrants are people who move to a new country. Study the map, and answer the questions.

German Immigration

1. Which states had the most German immigrants?

   _____

2. The Midwest is largely rural. What might German immigrants have done for a living?

   _____

3. The Midwest is far from the coasts. Immigrants had to travel by boat and then over land. The trip could take months. What do you think the trip was like?

   _____

   _____

WEEK 9 DAY 2

Name: _____ Date: _____

Directions: So many German immigrants settled in part of the Midwest that it is called the German Triangle. Follow the steps to draw it on the map.

German Immigration

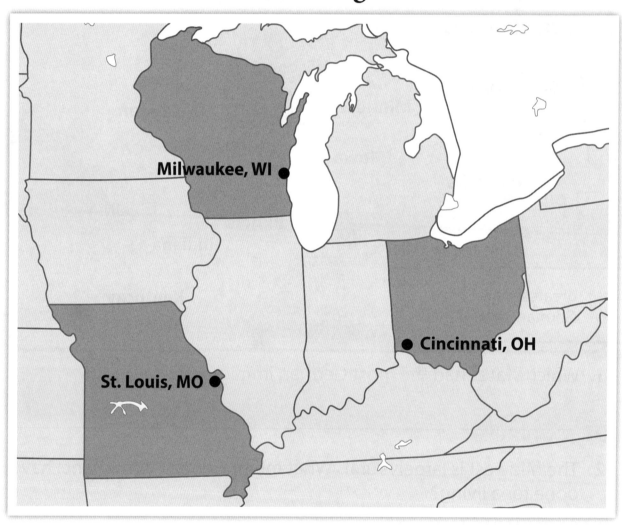

1. Draw a line from Milwaukee, Wisconsin, to Cincinnati, Ohio.

2. Draw a line from Cincinnati to St. Louis, Missouri.

3. Draw a line from St. Louis to Milwaukee.

4. Shade the triangle.

Name: _____ **Date:** _____

Directions: Read the text. Then, answer the questions.

Bringing German Culture

Many German people came to the United States in the 1800s. They brought their customs. Customs are ways of doing things. All cultures have their own customs. Germans put up trees for Christmas. Now, many cultures do this. Germans brought their own ideas, too. They started the first kindergarten. They also put PE in the school day.

Germans brought their language, too. Many of the words we use come from German. Words such as *noodle* and *hamburger* are German. You might use German words without knowing it!

1. What is a custom that German immigrants brought with them?

   _____

   _____

2. How would your school be different today without German ideas?

   _____

   _____

   _____

3. What are two words we use today that came from German?

   _____

   _____

WEEK 9 DAY 4

Name: _____   Date: _____

Directions: Study the photo, and read the caption. Then, answer the questions.

This is a Conestoga wagon. These wagons were designed by German settlers. They were used to move crops from rural areas to cities.

1. How many horses pull this Conestoga wagon?

   _____

2. There is no seat on the front of the wagon. Where do you think the driver rode?

   _____

   _____

3. The ends of these wagons were high and closed off. Why might that be good on rough, bumpy roads?

   _____

   _____

   _____

WEEK 9 DAY 5

Name: _____ **Date:** _____

Directions: Many words we use today come from the German language. Study the list. Put a checkmark next to words you have heard. Then, write a sentence using two of the words.

German Words	I Have Heard This Before
bratwurst	
delicatessen	
hamburger	
noodle	
pretzel	
pumpernickel	
sauerkraut	
kindergarten	

_____

_____

_____

Geography and Me

WEEK 10 DAY 1

Name: _____ Date: _____

Directions: A coastal plain is a flat area next to an ocean. Study the map, and answer the questions.

Atlantic and Gulf Coastal Plain

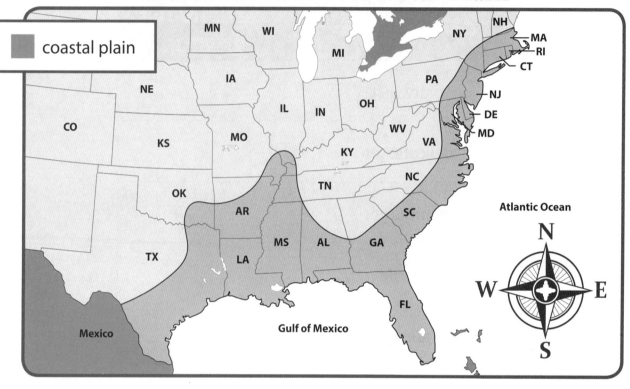

1. What ocean is east of the coastal plain?

   _____

2. What body of water is south of the coastal plain?

   _____

3. List six states in the coastal plain.

   _____   _____   _____

   _____   _____   _____

Name: _____ **Date:** _____

Directions: Follow the steps to complete the map.

1. Color the coastal plain yellow.

2. Color the rest of the United States green.

3. Color the Atlantic Ocean and the Gulf of Mexico blue.

4. Color Mexico orange.

5. Complete the legend to show what the colors mean.

WEEK 10 DAY 3

Name: _____ **Date:** _____

Directions: Read the text, and study the photo. Then, answer the questions.

The Atlantic and Gulf Coastal Plain

A coastal plain is a flat area of land next to an ocean. More than 20 states make up the Atlantic and Gulf Coastal Plain. The Atlantic Ocean is east of it. The Gulf of Mexico is south of it.

There are many resources in this region. There is coal, oil, natural gas, and timber. There are large ports, such as Baltimore, Maryland. A port is where ships pick up and drop off goods to sell.

People who work in the region farm and fish. Visitors swim, camp, and enjoy the beaches.

1. What is a coastal plain?

   _____

   _____

2. About how many states are in the Atlantic and Gulf Coastal Plain?

   _____

3. Why might ports be important in this region?

   _____

   _____

Name: _____ Date: _____

Directions: This photo shows a cargo ship in the port of Baltimore, Maryland. Use the photo to answer the questions.

1. What is a port?

   _____

   _____

2. How are goods being loaded onto the ship?

   _____

   _____

3. Why would companies want to bring their products to the port of Baltimore?

   _____

   _____

WEEK 10 DAY 5

Geography and Me

Name: _____ **Date:** _____

Directions: There are many ways to have fun in the coastal plain region. You can camp, kayak, bike, hike, and fish. Draw and label a picture of yourself visiting the coastal plain region.

Name: _____   Date: _____

Directions: These maps show two U.S. states. Study the maps. Then, answer the questions.

population: 2,059,000
17 people per square mile

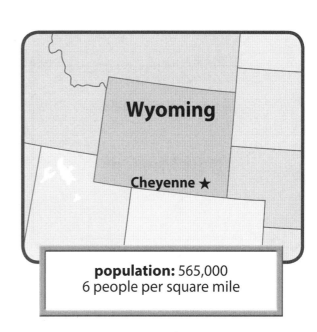

population: 565,000
6 people per square mile

1. Which state has more people?

   _____

2. What is the capital of Wyoming?

   _____

3. What is the capital of New Mexico?

   _____

4. Describe where each capital city is in its state.

   _____

   _____

   _____

WEEK 11
DAY 2

Name: _____ Date: _____

Directions: Study the maps, and follow the steps.

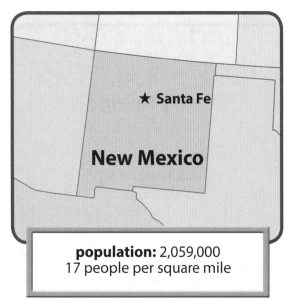

population: 2,059,000
17 people per square mile

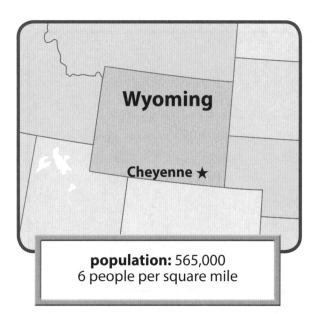

population: 565,000
6 people per square mile

The empty boxes each stand for one square mile.

New Mexico	Wyoming

1. Draw the number of people in one square mile in New Mexico.

2. Draw the number of people in one square mile in Wyoming.

3. Which state is more crowded? How do you know?

_____

_____

66   28623—180 Days of Geography   © Shell Education

Name: _____   Date: _____

Directions: Read the text. Then, answer the questions.

Population

Population is the number of people that live in a place. There are places where many people live close together. There are also places where few people live. There are big cities and small towns. People are very spread out in rural places.

People move from place to place for many reasons. They might want to be close to work. They might want to be close to family. They might need a house with more space or more land. Sometimes, people move for different weather. People may be tired of crowds in a big city. They might move to a rural place to have more room. People move to cities to be closer to jobs, schools, and shops. As people move, the populations change. Some areas get bigger. Other areas get smaller.

1. What happens to the population of a place if more people move there?

   _____

   _____

   _____

2. List three reasons why people might move.

   _____

   _____

   _____

WEEK 11 DAY 4

Name: _____   Date: _____

Directions: Study the table. Use the information to answer the questions.

State	Population	Size of the State
Wyoming	565,000	97,000 sq. mi (251,000 sq. km)
Ohio	11,537,000	45,000 sq. mi (117,000 sq. km)
New Mexico	2,059,000	122,000 sq. mi (316,000 sq. km)

1. Which state is the largest in size?

   _____

2. Which state has the most people?

   _____

3. Would you rather live in a large state with few people or a small state with many people? Why?

   _____

   _____

   _____

4. Ohio has over 250 people per square mile. New Mexico has 17 people per square mile. Wyoming only has 6. What does that say about Ohio?

   _____

   _____

68   28623—180 Days of Geography   © Shell Education

WEEK 11 DAY 5

Name: _____   Date: _____

Directions: Population is the number of people in an area. Draw a map of your classroom. Then, answer the questions.

Geography and Me

1. What is the population of your classroom?

   _____

2. How could you determine the population of your school?

   _____

   _____

WEEK 12 DAY 1

Name: _____ Date: _____

Directions: Study the map. Then, answer the questions.

American Rivers

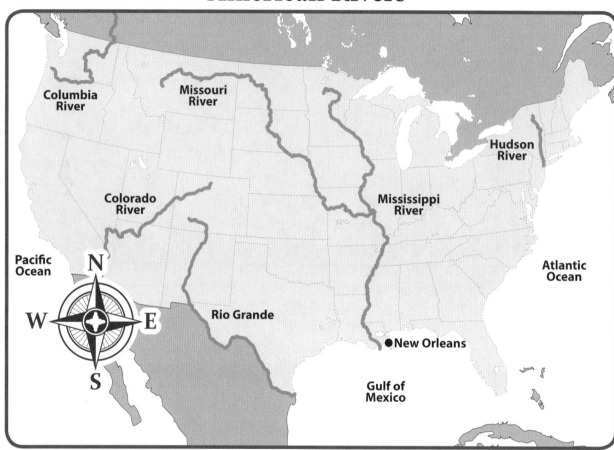

1. A delta is where a river flows into a large body of water. What is the city at the delta of the Mississippi River?

   _____

2. What body of water does the Mississippi River flow into?

   _____

3. What direction does the Mississippi River flow?

   _____

Name: _____ Date: _____

Directions: Study the map. Then, follow the steps.

American Rivers

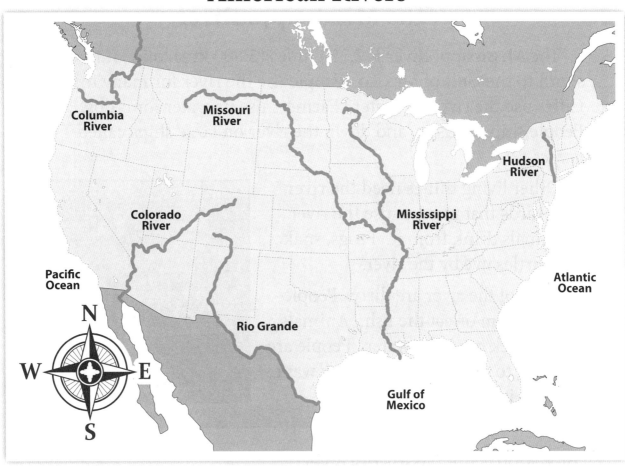

1. Trace the path of the Mississippi River in blue.

2. Place a star on the beginning of the Mississippi River.

3. Place an *X* where the Mississippi River empties into the Gulf of Mexico.

4. Label the Atlantic Ocean, the Pacific Ocean, and the Gulf of Mexico.

Challenge: Cover the names of the rivers with sticky notes. Label as many rivers as you can. Then, remove the sticky notes to check your answers.

WEEK 12 DAY 3

Name: _____ Date: _____

Directions: Read the text, and study the photo. Then, answer the questions.

The Mississippi River

The Mississippi River is 2,350 miles (3,800 km) long. It flows south to the Gulf of Mexico. People use the river for many things. Cities use the river for water. Farmers use its water for crops. People ship goods up and down the river on large ships called *barges*.

Other living things need the river, too. Birds that migrate use the river. Mammals drink from it. Frogs, snakes, and turtles live by the river.

Parts of the river are dirty. People cannot swim or eat the fish. Animals can get sick from the water. People are working to clean the river. They want to make it safe for all.

1. How long is the Mississippi River?

   _____

2. How is the river important to living things?

   _____

   _____

3. Parts of the river are dirty. What effect does this have?

   _____

   _____

Name: _____ Date: _____

Directions: Study the photos, and read the captions. Then, answer the questions.

This is New Orleans today.

This is New Orleans in 1851.

1. How has the city changed over time?

   _____

   _____

2. How has the port changed?

   _____

   _____

3. How have the ships on the Mississippi River changed?

   _____

   _____

WEEK 12 DAY 5

Name: _____ Date: _____

Directions: Many people are working to clean the Mississippi River. Create a poster to convince people that clean water is important to our world.

Geography and Me

Name: _____ Date: _____

Directions: Study the map, and follow the steps.

The Great Lakes

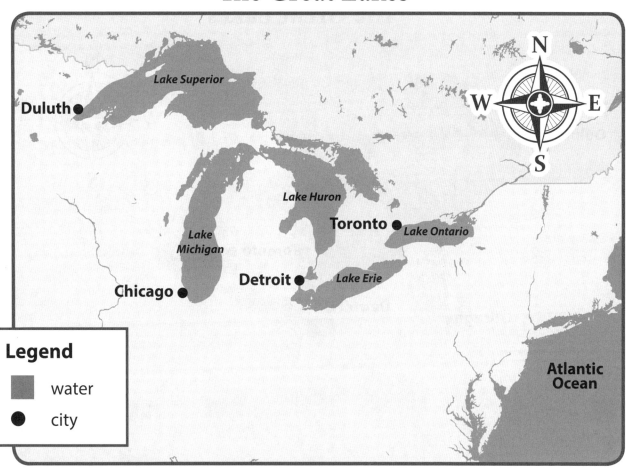

1. Find Duluth, Minnesota, on the map. Which Great Lake is Duluth near?

   _____

2. Start at Duluth, Minnesota. Use a blue crayon to trace a path through the Great Lakes to Toronto.

3. How many of the Great Lakes does your path go through?

   _____

WEEK 13 DAY 2

Name: _____ Date: _____

Directions: This is a map of the Great Lakes. The lakes are connected and lead to the Atlantic Ocean. Study the map, and answer the questions.

The Great Lakes

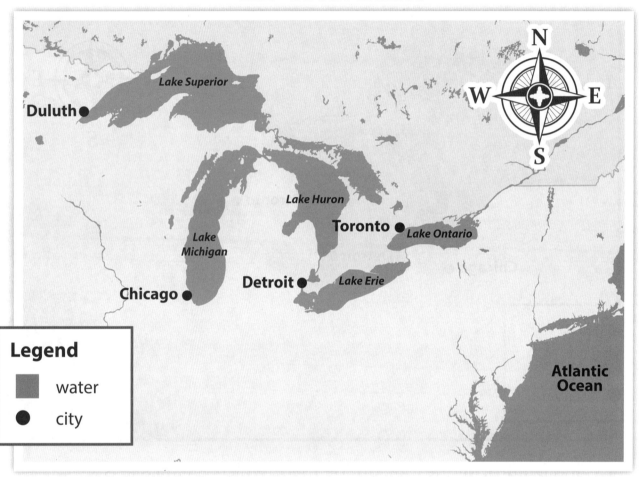

1. Draw a box around the title of the map.

2. Circle the compass rose.

3. Put a star next to the legend.

4. Find one item from the legend on the map. Draw a cloud around it. What did you find?

_____

_____

Name: _____ **Date:** _____

Directions: Read the text, and study the photo. Then, answer the questions.

The Great Lakes

The Great Lakes are a set of five large bodies of water. They were carved out by glaciers long ago. Glaciers are large sheets of ice that cover the land. The St. Lawrence Seaway connects the Great Lakes to the Atlantic Ocean. The lakes give water to 48 million people.

Not much water leaves the lakes. When pollution gets in the lakes, it stays. Pollution comes from farms, cities, and factories. It may not be safe to eat the fish. Birds that eat the fish may not have healthy babies. People are working to keep the Great Lakes clean.

1. How many lakes are in the Great Lakes?

   _____

2. How could you travel from the Atlantic Ocean to the Great Lakes?

   _____

   _____

3. Why is it important to keep the Great Lakes clean?

   _____

   _____

   _____

WEEK 13 DAY 4

Name: _____ **Date:** _____

Directions: This table shows how deep each lake is overall. Study the table. Then, answer the questions.

Great Lake	Average Depth
Superior	489 ft. (149 m)
Michigan	279 ft. (85 m)
Huron	185 ft. (59 m)
Erie	62 ft. (19 m)
Ontario	283 ft. (86 m)

1. Which lake is the shallowest?

   _____

2. Which lake is the deepest?

   _____

3. How much deeper is Lake Michigan than Lake Huron? How do you know?

   _____

   _____

4. Lake Superior holds the most water. Does the chart support this fact? If so, how?

   _____

   _____

   _____

Name: _____ **Date:** _____

Directions: What is the closest body of water where you live? Draw and label that body of water.

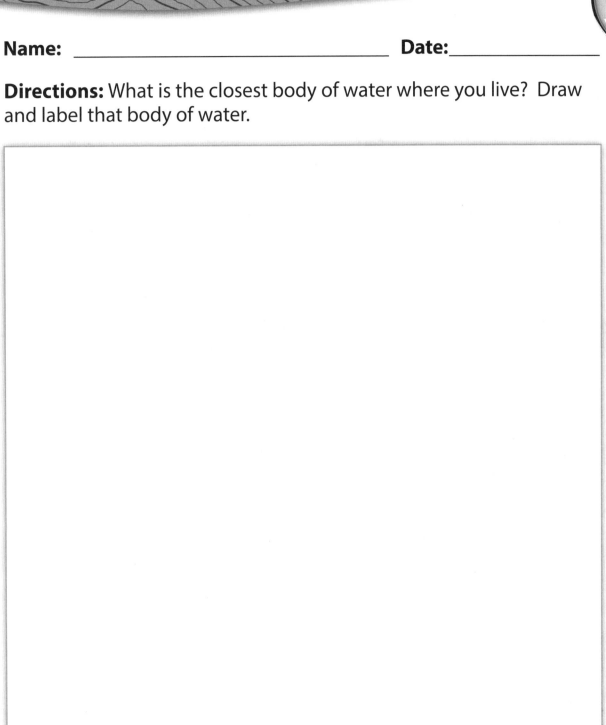

Geography and Me

WEEK 13
DAY 5

Name: _____ **Date:** _____

Directions: Study the map. Answer the questions.

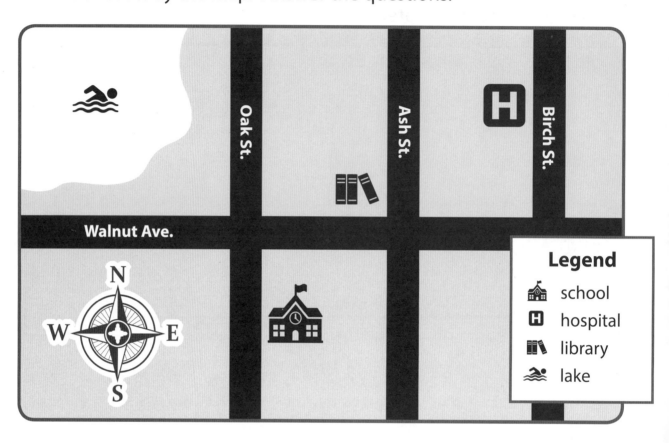

1. What directions is the hospital from the library?

   _____

2. What directions is the school from the lake?

   _____

3. On which street is the school?

   _____

4. What location is closest to the school?

   _____

Name: _____ **Date:** _____

Directions: Maps can have grids to help people find places. Use the grid to follow the steps.

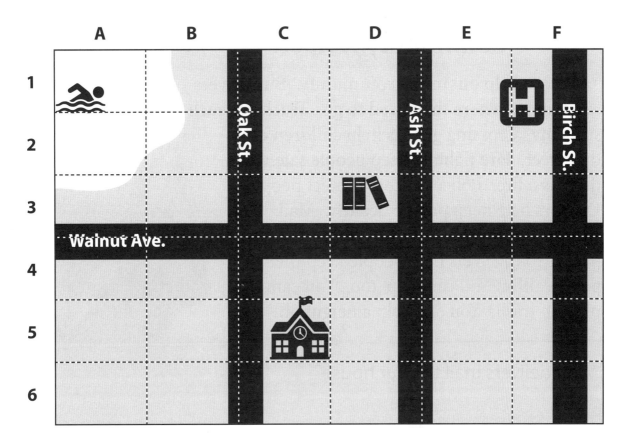

1. In which box is the library?

   _____

2. Draw a house symbol in the box at D, 6.

3. Draw a store symbol in the box at A, 6.

4. Draw a fire station symbol in the box at E, 3.

5. Draw a baseball field symbol in the box at C, 1.

WEEK 14 DAY 3

Name: _____ **Date:** _____

Directions: Read the text, and study the photo. Then, answer the questions.

Community Helpers

People help out in the community. Bus drivers take people where they need to go. Teachers teach students. Crossing guards help children cross the street. Fire fighters keep people safe when there is a fire. Police officers protect people. Doctors help keep people healthy. Mail carriers bring mail. Farmers grow food for people to eat. Vets help people keep their pets healthy. You can help, too. You can pick up trash. You can help a neighbor.

1. Who delivers mail to your house?

   _____

2. Who helps students get to school?

   _____

3. Who can help people in an emergency?

   _____

4. The text says that you can help by picking up trash or helping a neighbor. How else could you help out in your community?

   _____

   _____

Name: _____ **Date:** _____

Directions: Every day, students at a school go home in different ways. Use the table to answer the questions.

How Students Go Home	Number of Students
walking	4
riding in a day care van	3
riding in a car	6
riding a bike	1
riding a bus	10

1. What is the most common way students go home?

   _____

2. What is the least common way students go home?

   _____

3. How many more students ride a bus than walk home? How do you know?

   _____

   _____

   _____

4. Which way from the table would you rather take to get home? Why?

   _____

   _____

   _____

WEEK 14 DAY 5

Geography and Me

Name: _____   Date: _____

Directions: Communities are where people live, work, and play. Draw pictures of you where you live, work, and play. Hint: You work hard at school.

Where I Live

Where I Work

Where I Play

WEEK 15 DAY 1

Name: _____ Date: _____

Directions: Rivers and lakes are important sources of freshwater. The ocean contains saltwater. Study the map, and answer the questions.

American Lakes and Rivers

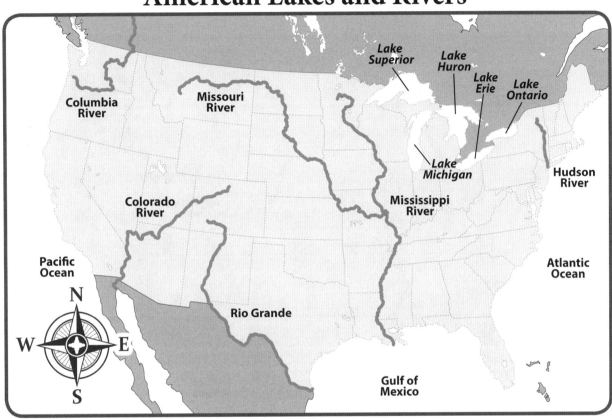

1. Which river flows into the Mississippi River?

   _____

2. Which river is east of the Mississippi River?

   _____

3. Can people drink water from the ocean? Why or why not?

   _____

   _____

   _____

WEEK 15 DAY 2

Name: _____   Date: _____

Directions: Follow the steps to complete the map.

American Lakes and Rivers

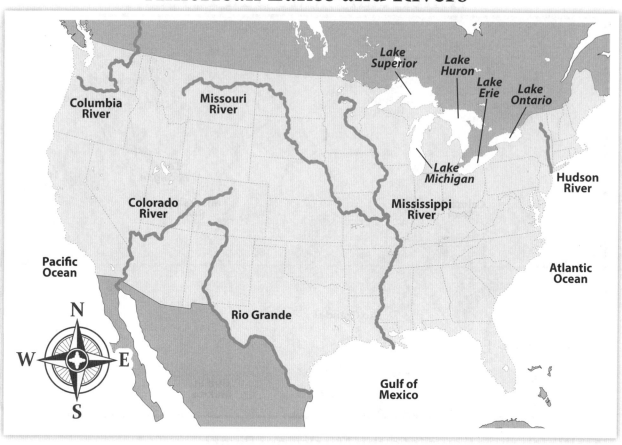

1. Color the Great Lakes blue.

2. Trace the rivers in blue.

3. Use a different color to shade the oceans.

4. Label the Pacific Ocean in the west.

5. Label the Atlantic Ocean in the east.

6. Label the Gulf of Mexico in the south.

7. Draw stars where the rivers empty into the oceans.

Name: _____ **Date:** _____

Directions: Read the text. Then, answer the questions.

Clean Water

Three quarters of Earth is covered in water. But, most of that is saltwater. Only a tiny bit of it is freshwater. Only half of that is easy to get and use. Some of it is frozen in glaciers.

Clean water is important. You can help keep water clean. You can clean trash out of streams. You can put plants next to streams. This helps stop erosion. Recycle paper, plastic, and cans. Pick up litter. Take shorter showers. We can all help protect our water.

1. What can you do to keep water clean?

   _____

   _____

   _____

2. Why can't we use all types of water on the planet?

   _____

   _____

   _____

3. Why is it important to have clean water?

   _____

   _____

   _____

WEEK 15 DAY 4

Name: _____   Date: _____

Directions: Study the picture. Then, answer the questions.

1. What kinds of trash do you see in the river?

   _____

   _____

2. What could you do to keep rivers clean?

   _____

   _____

   _____

3. How might litter end up in the river?

   _____

   _____

WEEK 15 DAY 5

Name: _____ Date: _____

Directions: Make a list of all the times you use water in one day. Then, draw a picture to show how you can help protect our freshwater.

Geography and Me

WEEK 16 DAY 1

Name: _____ Date: _____

Directions: Study the map, and answer the questions.

1. How many cities are shown on this map?

   _____

2. Which direction would you drive to go from Las Vegas to Lake Mead?

   _____

3. Which direction would you drive to go from Carson City to Lake Tahoe?

   _____

Name: _____ Date: _____

Directions: Follow the steps, and answer the questions.

1. Outline the lakes in blue.

2. Add a blue square to the legend to show that lakes are blue.

3. Add a black dot to the legend to show that dots stand for cities.

WEEK 16 DAY 3

Name: _____ Date: _____

Directions: Read the text. Then, answer the questions.

Always Changing

The way we live is always changing. New inventions can make our lives easier. Electric lights help people work after dark. Paved roads make travel easier. Air conditioning makes hot places more comfortable. Cars let us live farther from work. The Internet helps us share information. We can share new ideas right away.

The land around us can change, too. New homes and schools can be built. New roads can be built, and cities can grow. As people move to the suburbs, those towns grow, too. People may move to rural places. Or rural places can turn into towns. But, more people means less land for animals. It can also mean more pollution. As communities grow, they use more resources. They use wood for houses, gas for cars, and clean water. We must be careful about how we use the land.

1. How have inventions changed the way we live?

   _____

   _____

2. How can the land around us change?

   _____

   _____

3. Name one resource that communities use.

   _____

Name: _____ Date: _____

Directions: Study the photos closely. Then, answer the questions.

This is Las Vegas, Nevada, in 1910.

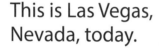

This is Las Vegas, Nevada, today.

1. In 1910, was Las Las Vegas rural, suburban, or urban?

   _____

2. Is Las Vegas rural, suburban, or urban today?

   _____

3. How has the city changed?

   _____

   _____

   _____

WEEK 16 DAY 5

Name: _____ Date: _____

Directions: What do you think your community looked like long ago? Complete the Venn diagram to compare and contrast your community long ago and today.

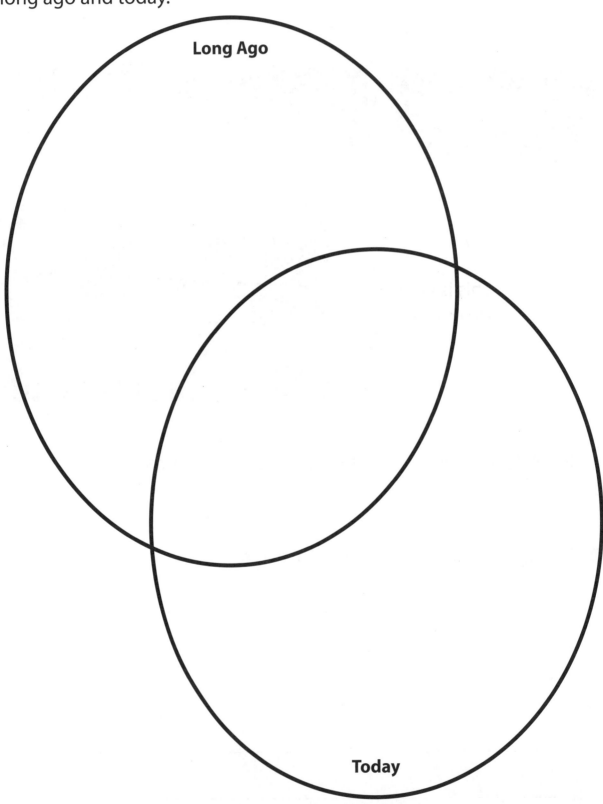

Name: _____ Date: _____

Directions: A canal is a waterway that ships can pass through. The Panama Canal lets ships pass through Panama. Study the map, and answer the questions.

1. Which ocean is north of the Panama Canal?

   _____

2. Which ocean is south of the Panama Canal?

   _____

3. What continents does Panama connect?

   _____

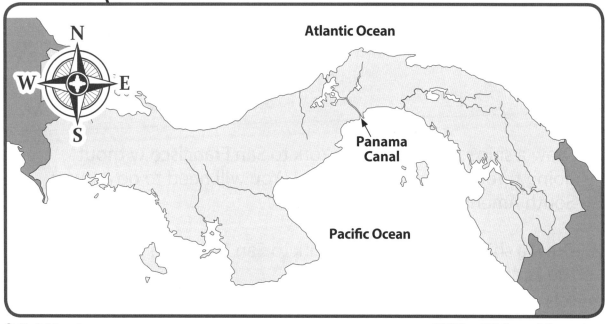

Name: _____ **Date:** _____

Directions: The Panama Canal was completed in 1914. It made the trip from New York to San Francisco five months shorter. Follow the steps to show the routes.

1. Draw a ship's path from New York to San Francisco without going through the Panama Canal. You will need to go around South America.

2. Draw a ship's path from New York to San Francisco going through the Panama Canal.

Name: _____ **Date:** _____

Directions: Read the text, and study the photo. Then, answer the questions.

The Panama Canal

The Panama Canal is a waterway. It connects the Atlantic and the Pacific Oceans. It was built by the United States. People wanted a faster route to the West. The canal took 10 years to build. It is 50 miles (80 km) long. Larger canals have been built. But this one was hard to build. Workers had to cut through mountains. The weather was bad much of the time. Many workers got sick, too.

The canal changed how ships travel. Before, ships had to sail around South America. Now, they have a shorter route. This saves time and money. Over 12,000 ships use the canal each year.

1. How long is the Panama Canal?

   _____

2. Why was the Panama Canal hard to build?

   _____

   _____

3. How did the Panama Canal change travel?

   _____

   _____

WEEK 17 DAY 4

Name: _____ Date: _____

Directions: Look at the map, and study the table. Then, answer the questions.

Distance by Ship from San Francisco to New York	
before the Panama Canal	14,000 miles (22,500 km)
after the Panama Canal	4,800 miles (7,700 km)

1. How long was the trip from New York to San Francisco before the Panama Canal?

   _____

2. How long is the trip after the Panama Canal?

   _____

3. How many more miles was the trip before the Panama Canal?

   _____

Name: _____ **Date:** _____

Directions: Imagine you could make a shortcut to speed up travel. Answer the questions, and follow the steps to show your shortcut.

1. Where would you make your shortcut?

   _____

   _____

2. Why would you make your shortcut there?

   _____

   _____

   _____

3. Draw a map of the long way. Then, draw and label your shortcut.

WEEK 18 DAY 1

Name: _____ Date: _____

Directions: Niagara Falls is made up of three large waterfalls. Many people visit the falls every year. Study the map, and answer the questions.

1. In which two countries is Niagara Falls?

   _____

2. In which U.S. state is Niagara Falls?

   _____

3. In what Canadian province is Niagara Falls?

   _____

Name: _____ Date: _____

Directions: Follow the steps to label the map.

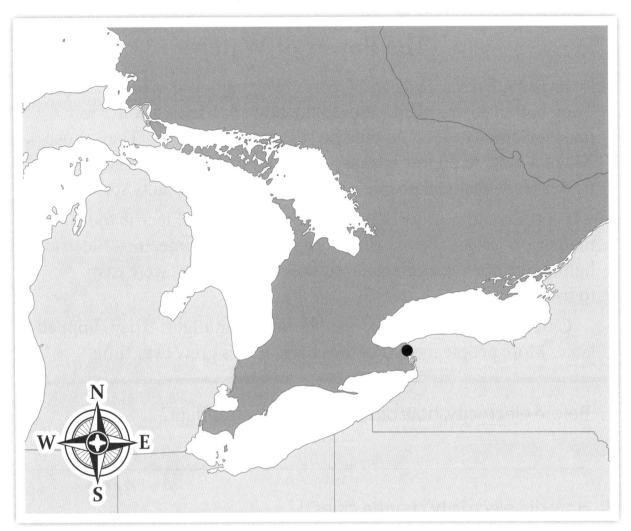

1. Label Ontario, Canada.

2. Label New York State.

3. Label Niagara Falls.

4. Draw a waterfall symbol near Niagara Falls.

5. Label Lake Ontario. It is south of Niagara Falls.

6. Label Lake Erie. It is south of Niagara Falls.

WEEK 18 DAY 3

Name: _____ Date: _____

Directions: Read the text. Then, answer the questions.

The Power of Water

In 1895, Nikola Tesla and George Westinghouse made history. They built a power plant. It used the power of falling water to make electricity. It could then be sent to other places. They chose Niagara Falls to do this because the water has so much force. Their power plant changed people's lives. More power plants sprung up.

People could now work more easily after dark. Electric tools made work easier. People did more work in less time, too. Electric lights were safer. Candles started fires. These lights were easy to use.

Cities changed over time. People worked at night. They shopped later. More people moved to the cities. Cities grew over time.

1. Before electricity, how did people work after dark?

_____

2. How did electricity change cities?

_____

_____

3. How did electricity change people's lives?

_____

_____

_____

28623—180 Days of Geography © Shell Education

Name: _____ **Date:** _____

Directions: Easy access to electricity changed cities. Cities grew bigger. Buildings could be taller with electric elevators. Study the pictures. Then, answer the questions.

This is New York City in 1880.

This is New York City today.

1. How is the city different today?

   _____

   _____

2. How are the buildings different today?

   _____

   _____

3. What impact did electricity have on New York City?

   _____

   _____

   _____

WEEK 18 DAY 5

Name: _____ Date: _____

Directions: Make a list of the things you do that use electricity. Then, answer the questions.

Geography and Me

1. How would your life be different without electricity?

2. What do you think someone from the 1800s would think about your life?

Name: _____ **Date:** _____

Directions: Study the map. Then, answer the questions.

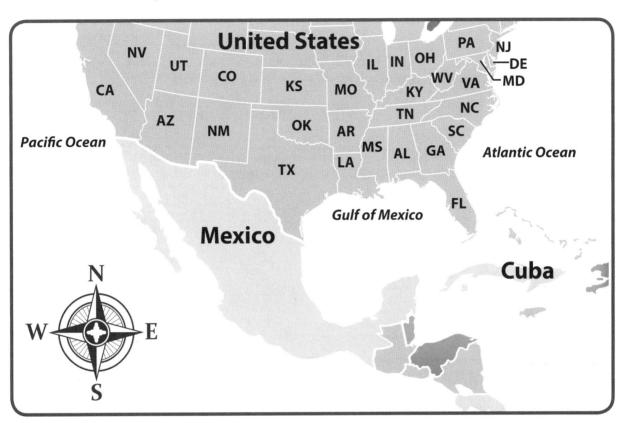

1. Name three countries that border the Gulf of Mexico.

   _____

   _____

2. Which U.S. states border the Gulf of Mexico?

   _____

   _____

3. Which direction would you go to sail from the Gulf of Mexico to the Atlantic Ocean?

   _____

Name: _____ **Date:** _____

Directions: Follow the steps to label the map.

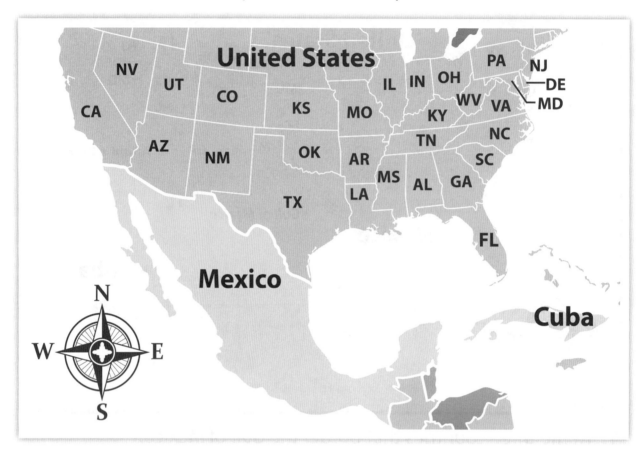

1. Label the Atlantic Ocean.

2. Label the Pacific Ocean.

3. Label the Gulf of Mexico.

4. Color the land that borders the Gulf of Mexico.

5. Draw a route a ship might take from southern Texas out of the Gulf of Mexico.

WEEK 19 DAY 3

Name: _____ **Date:** _____

Directions: Read the text. Then, answer the questions.

The Gulf of Mexico

A gulf is a part of the sea. It is almost enclosed by land. The Gulf of Mexico is the largest gulf in the world.

The Gulf of Mexico is important to people. People use resources from the gulf. They get seafood, such as shrimp and crabs. They get oil from the gulf. Ships take cargo through the gulf.

Animals need the gulf, too. Many marine mammals, such as dolphins, whales, and manatees live in the gulf. Sharks and sea turtles also call the gulf home. Some of these animals are threatened or endangered. This means there are few of them left.

The Gulf of Mexico needs protection. Many river systems flow into it. These rivers bring pollution. It can hurt plants and animals that live in the gulf. When people eat these things, they may get sick. People work hard to keep the gulf clean.

Read About It

1. What is a gulf?

   _____

2. Why is the Gulf of Mexico important to people?

   _____

   _____

3. What might happen if people continue to pollute the Gulf of Mexico?

   _____

   _____

WEEK 19 DAY 4

Think About It

Name: _____ Date: _____

Directions: Study the list of threatened and endangered species in the Gulf of Mexico. Then, answer the questions.

Sea Turtle Species	Status
green sea turtle	threatened
hawksbill sea turtle	endangered
Kemp's ridley sea turtle	endangered
leatherback sea turtle	endangered
loggerhead sea turtle	threatened

Marine Mammals	Status
fin whale	endangered
sei whale	endangered
sperm whale	endangered

1. How many types of whales in the Gulf of Mexico are endangered?

   _____

2. How many types of sea turtles in the Gulf of Mexico are threatened?

   _____

3. Endangered animals are at higher risk of dying out than threatened animals. What does the chart tell you about how much danger these animals are in?

   _____

   _____

4. What will happen to an endangered species if the last one dies?

   _____

   _____

28623—180 Days of Geography   © Shell Education

Name: _____ **Date:** _____

Directions: Draw a poster convincing people to save the animals in the Gulf of Mexico.

WEEK 19 DAY 5

Geography and Me

WEEK 20 DAY 1

Reading Maps

Name: _____ Date: _____

Directions: Study the map. Then, answer the questions.

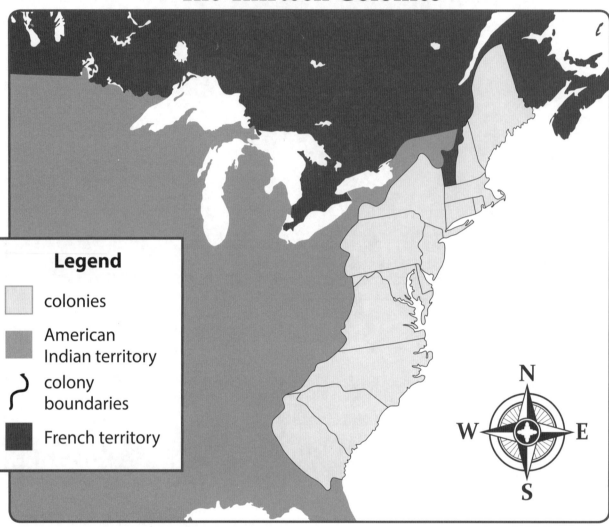

1. What is the title of the map?

   _____

2. Circle the compass rose.

3. What is listed in the legend?

   _____

   _____

Name: _____ Date: _____

Directions: Study the map. Then, follow the directions.

The Thirteen Colonies

1. Outline the colonies in red.

2. Color the northernmost colony blue.

3. Color the southernmost colony green.

WEEK 20 DAY 3

Name: _____ Date: _____

Directions: Read the text. Then, answer the questions.

Colonies in the New World

In the 1600s, people from Europe began coming to the New World. Rulers in Europe claimed the land. They formed colonies. Many people came to live there. They built towns and roads. But, American Indians already lived on the land. At first, the two groups tried to get along. They traded fur, tools, and food. Yet, fights often broke out. The settlers spread diseases, too. The tribes had never had these diseases before. Many of their people died.

Over time, the settlers moved west. They killed many bison. This was a food source for many tribes. The settlers took more tribal land. Their lives changed forever.

1. What did the American Indians trade with the settlers?

   _____

   _____

2. Why do you think fights broke out between the two groups?

   _____

   _____

3. What was the effect of the settlers on the American Indians' lives?

   _____

   _____

   _____

Name: _____ **Date:** _____

Directions: Bison roamed North America before European settlers came. When the settlers moved west, they killed many bison. Today, bison are protected. Read the table, and answer the questions.

Time	Number of Bison
1600 (before settlers)	20–30 million
1889	1,091
today	500,000

1. How many bison roamed North America before the settlers?

   _____

2. How many bison are in North America now?

   _____

3. Do you think making bison a protected species made a difference? How do you know?

   _____

   _____

   _____

4. How many bison do you think there will be in 2050? Why?

   _____

   _____

   _____

WEEK 20 DAY 5

Name: _____ Date: _____

Directions: Draw yourself visiting a prairie where bison live.

Name: _____ Date: _____

Directions: The United States can be divided into regions. A region is a place where the land or weather is different from the places around it. Study the map, and answer the questions.

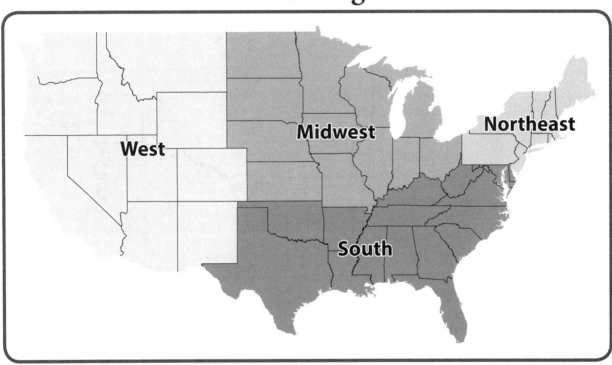

1. How many different regions are on this map?

   _____

2. Name the regions shown on this map.

   _____

   _____

3. What do the names of the regions tell you?

   _____

   _____

WEEK 21
DAY 2

Name: _____ **Date:** _____

Directions: Follow the steps to complete the map.

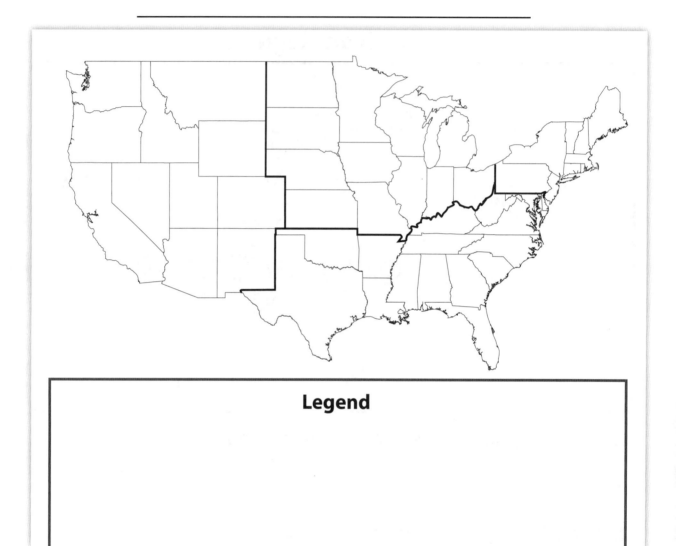

1. Add a compass rose to the map.

2. Label the regions: Northeast, South, Midwest, and West.

3. Shade each region a different color. Complete the legend to explain what the colors mean.

4. Add a title to the map.

Name: _____ **Date:** _____

Directions: Read the text. Then, answer the questions.

Cooling the South and West

The southern region is hot in summer. Parts of the western region are hot, too. In 1902, Willis Carrier invented air conditioning. People added it to their homes. They put it in businesses. Buildings in hot places could be kept cool. People could live and work in warmer places. The number of people in the South and the West rose.

Air conditioning changed where and how people lived. Before, they opened their windows to cool the house. People sat on the porch to cool off. Now, people stay inside when it is hot. They can make it as warm or as cool as they want.

1. When was air conditioning invented?

   _____

2. How did air conditioning change where people lived?

   _____

   _____

   _____

3. How did air conditioning change how people lived?

   _____

   _____

   _____

WEEK 21 DAY 4

Name: _____ **Date:** _____

Directions: Air conditioning became common in the 1950s. People moved to hotter places, such as the South and the West. Study the table. Then, answer the questions.

Largest Cities in 1950	Region	Largest Cities in 2010	Region
New York, New York	Northeast	New York, New York	Northeast
Chicago, Illinois	Midwest	Los Angeles, California	West
Philadelphia, Pennsylvania	Northeast	Chicago, Illinois	Midwest
Los Angeles, California	West	Houston, Texas	South
Detroit, Michigan	Midwest	Philadelphia, Pennsylvania	Northeast
Baltimore, Maryland	South	Phoenix, Arizona	West
Cleveland, Ohio	Midwest	San Antonio, Texas	South
St. Louis, Missouri	Midwest	San Diego, California	West
Washington, DC	South	Dallas, Texas	South
Boston, Massachusetts	Northeast	San Jose, California	West

1. How many of the largest cities were in the South or West in 1950?

   _____

2. How many of the largest cities were in the South or West in 2010?

   _____

3. How did air conditioning affect the growth of cities?

   _____

   _____

   _____

WEEK 21 DAY 5

Name: _____ **Date:** _____

Directions: Create a map of your classroom. Label separate regions of your classroom. For example, you could label the classroom library or the teacher's desk.

Geography and Me

WEEK 22 DAY 1

Name: _____ Date: _____

Directions: The Continental Divide is made up of different mountain ranges. Study the map. Then, answer the questions.

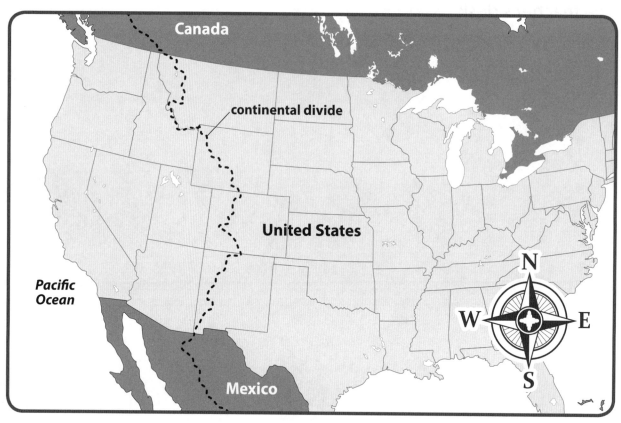

1. What three countries does the Continental Divide go through?

   _____

2. What ocean is west of the Continental Divide?

   _____

3. Does the Continental Divide run mostly north and south or east and west? How do you know?

   _____

   _____

   _____

Name: _____ **Date:** _____

Directions: The Continental Divide separates rivers. Rivers on the east side flow to the Atlantic Ocean. Rivers on the west side flow to the Pacific Ocean. Study the map, and follow the steps.

1. Color the land east of the Continental Divide blue. Draw an arrow to show that the water moves east.

2. Color the land west of the Continental Divide red. Draw an arrow to show that the water moves west.

The Continental Divide

The Continental Divide passes through several mountain ranges. It starts in Alaska and goes to South America. It is called a *divide* because it splits the rivers. No rivers cross the divide. All the rivers on the east side flow east. All the rivers on the west side flow west.

When settlers moved west, they had to cross the divide. It was a tough journey. Many went through South Pass. It is not as steep there. That made it easier for wagons to cross. South Pass was part of the Oregon Trail. Many people used this trail to move west.

1. Why is it called a *continental divide*?

   _____

   _____

2. Where did settlers on the Oregon Trail cross the Continental Divide?

   _____

3. Look at the picture. What might have been difficult for a settler crossing the Continental Divide?

   _____

   _____

Name: _____ **Date:** _____

Directions: The trip across the Continental Divide took four months. This table shows some of the food a family would need on the Oregon Trail. Study the table. Then, answer the questions.

Food	Amount
flour	600 pounds (270 kg)
biscuits	120 pounds (54 kg)
bacon	400 pounds (180 kg)
tea	4 pounds (2 kg)
sugar	100 pounds (45 kg)
lard	200 pounds (91 kg)

1. How much more sugar than tea was needed?

   _____

2. How much more bacon than lard was needed?

   _____

3. Which item did the settlers need the most of?

   _____

4. Does anything in the table surprise you? Why or why not?

   _____

   _____

   _____

   _____

WEEK 22 DAY 5

Name: _____ Date: _____

Directions: In the 1800s, many families packed a covered wagon and headed west. Draw and write to show what you would pack in a wagon for the trip.

Geography and Me

_____
_____
_____
_____
_____
_____
_____

WEEK 23 DAY 1

Name: _____ Date: _____

Directions: Study the map. Then, answer the questions.

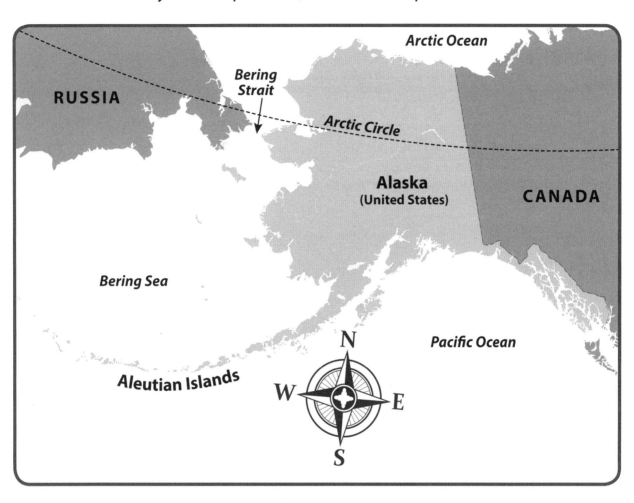

1. Alaska is part of the United States. What country borders Alaska to the east?

   _____

2. What island chain extends from southern Alaska?

   _____

3. What waterway separates Alaska from Russia?

   _____

WEEK 23 DAY 2

Name: _____ Date: _____

Directions: Follow the steps to color the map.

Creating Maps

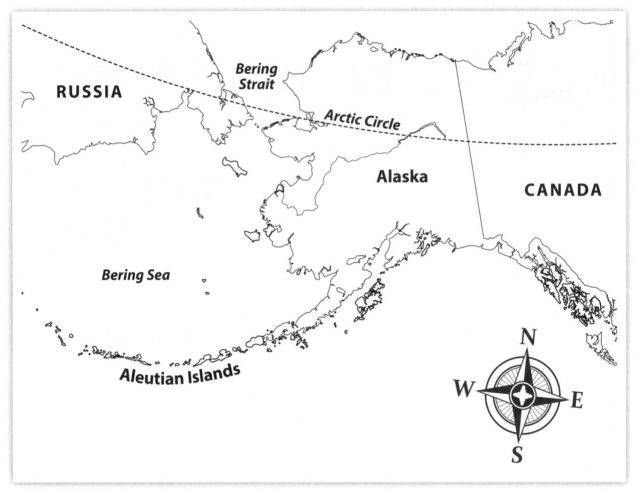

1. A peninsula is a piece of land that has water on three sides. Color the peninsulas in Alaska brown.

2. Color the Aleutian Islands orange.

3. The Arctic Circle passes through Alaska. On the map, it is a dashed line. Trace the line in purple.

4. Color Canada blue.

5. Color Russia green.

Name: _____ Date: _____

Directions: Read the text. Then, answer the questions.

Alaska

Alaska is very cold for much of the year. Some places only thaw for a short time. The Alaska Natives learned to live in these harsh climates. They hunted and fished. They used the land wisely.

Russians came to the region in 1750. They wanted to bring furs back home to sell. But they brought diseases with them. They spread them to many Alaska Natives.

The United States bought Alaska in 1867. It became a state in 1959. People built roads. More people came to live in the state. This made it harder for the Alaska Natives to find food.

Today, Alaska Natives use modern things. But they still value their old ways of life. They are trying to preserve their cultures.

1. How did the Alaska Natives live before the Russians came?

   _____

   _____

2. How did the Russians affect the Alaska Natives?

   _____

   _____

3. How do Alaska Natives live today?

   _____

   _____

WEEK 23 DAY 4

Name: _____ Date: _____

Directions: Alaska Natives used different types of transportation. Study the table. Then, answer the questions.

Type of Transportation	Use
umiak (small covered canoe)	hunting whales
baidarka (skin-covered kayak)	traveling long distances on water
sleds pulled by dogs	traveling over snow

1. What might an Alaska Native use to travel over water to trade with another village? How do you know?

   _____

   _____

2. What boat would an Alaska Native use to hunt whales?

   _____

3. How did Alaska Natives travel over frozen land?

   _____

4. Which type of transportation do you think was the fastest? Why?

   _____

   _____

   _____

WEEK 23 DAY 5

Name: _____ Date: _____

Directions: In Barrow, Alaska, the sun does not set for 80 straight days. Imagine you lived in Barrow. Draw a picture, and write a paragraph convincing your parents to let you stay up later in the summer.

Geography and Me

© Shell Education 28623—180 Days of Geography

Name: _____ **Date:** _____

Directions: This map shows the Erie Canal. It is a human-made waterway. It goes from the Hudson River to the Great Lakes. Use the map to answer the questions.

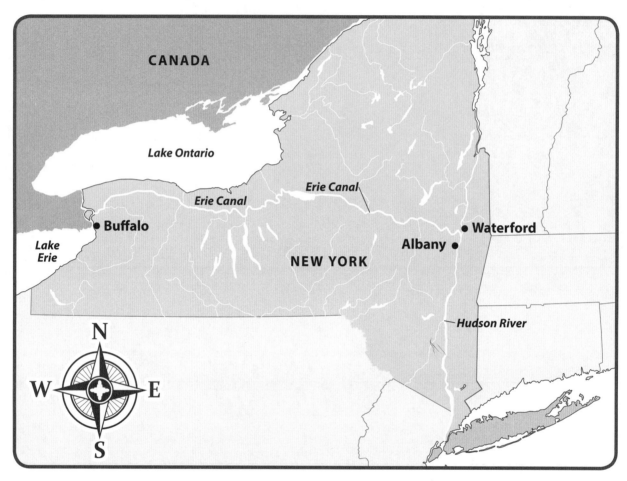

1. If you went from Buffalo to Albany, what direction would you be traveling?

   _____

2. Which two Great Lakes border New York?

   _____

3. What country is north of New York?

   _____

Name: _____ **Date:** _____

Directions: Follow the steps to complete the map.

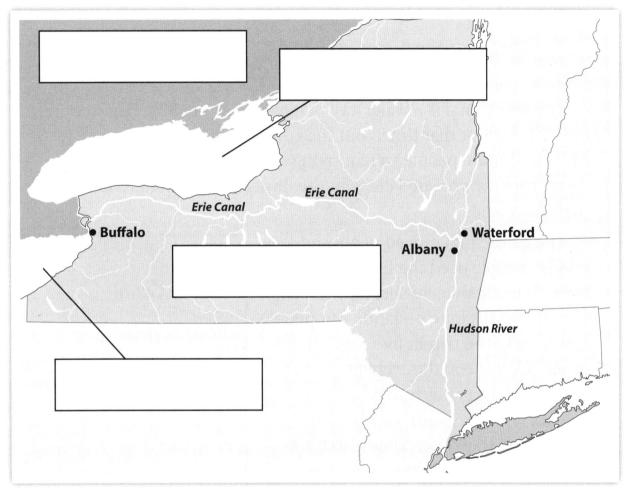

1. Label Canada and New York.

2. Label Lake Ontario. It is north of Lake Erie.

3. Label Lake Erie.

4. Trace the Hudson River in blue.

5. Trace the Erie Canal in red.

6. Add a compass rose to the map.

WEEK 24 DAY 3

Name: _____ **Date:** _____

Directions: Read the text, and study the map. Then, answer the questions.

Moving West

The Erie Canal is a waterway that was built in 1825. There were only 24 states at that time. But the country was growing quickly. More and more people were moving west. They wanted to get away from the crowded cities in the East. They wanted to farm their own land.

The Erie Canal moved people and goods. People used the Erie Canal to go west. It made the trip shorter. The trip was two weeks long on land. But people could make the trip in five days on the canal. Goods could travel between the Atlantic Ocean and the Midwest on the canal. Barges took goods made in factories to the West. They also took goods from farms to the East. This made trade easier and cheaper.

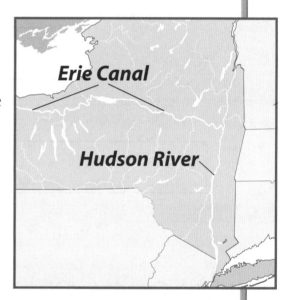

1. How did the Erie Canal help the United States grow?

   _____

   _____

2. What do you think people thought about the Erie Canal in 1825? Why?

   _____

   _____

Name: _____ **Date:** _____

Directions: This chart shows the time it took to travel from Albany, New York, to Buffalo, New York, in 1825.

Type of Travel	Length of Trip
stagecoach or wagon	14 days
boat on the Erie Canal	5 days

1. How did the Erie Canal change the trip from Albany to Buffalo?

   _____

   _____

2. How many more days did the trip take by wagon than by boat?

   _____

3. Why might people want the trip to be shorter?

   _____

   _____

4. How do you think the Erie Canal changed people's lives?

   _____

   _____

   _____

WEEK 24 DAY 4

Think About It

WEEK 24 DAY 5

Name: _____ **Date:** _____

Directions: Before the Erie Canal, people made many things at home. After the canal was built, people bought more things from stores. Write things your family makes and buys.

Make at Home	Buy in Stores

Geography and Me

Name: _____ Date: _____

Directions: Study the map. Then, answer the questions.

American Regions

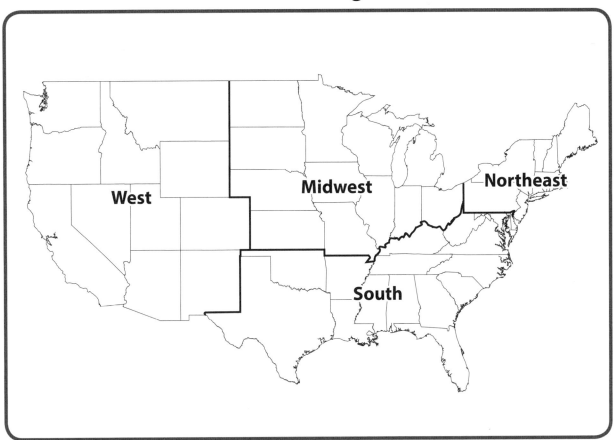

1. On this map, how many regions are there in the United States?

   _____

2. How many states are in the Midwest region?

   _____

3. Color all the states in the Midwest region orange.

4. Color the other three regions in three different colors.

WEEK 25 DAY 2

Name: _____ **Date:** _____

Directions: Follow the steps to complete the map.

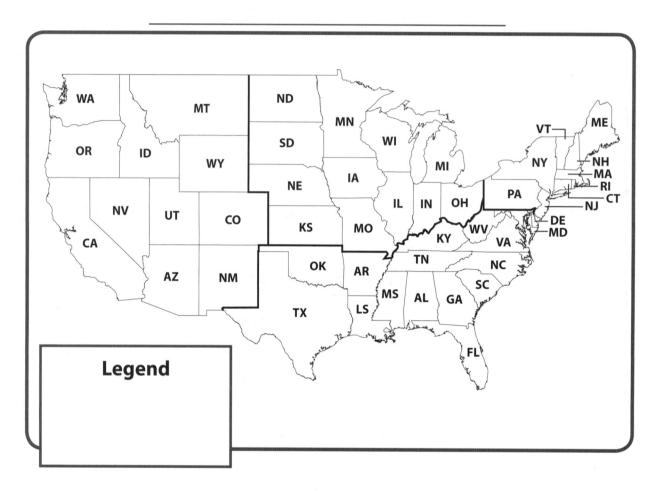

1. A lot of corn is grown in Indiana (IN), Illinois (IL), Iowa (IA), Missouri (MO), Nebraska (NE), and Kansas (KS). Add a symbol to the map showing that these states grow corn.

2. Complete the legend to show what the symbol means.

3. Add a title to the map.

4. Add a compass rose to the map.

28623—180 Days of Geography © Shell Education

Name: _____ **Date:** _____

Directions: Read the text, and study the photo. Then, answer the questions.

Growing Corn in the Midwest

The Midwest region is perfect for growing corn. The Midwest is called the Corn Belt. It has good soil, warm nights, and hot days. Corn takes a lot of water to grow. Yet it is the biggest crop grown in the United States.

Corn is food for farm animals. It can be made into fuel. Corn is also found in many of our foods. Corn syrup can be used as a sweetener. It can be found in chewing gum and ketchup. Corn is in taco shells and cereal. It is in snack foods and popcorn. Corn is also used in soaps and glue. Many people depend on this crop.

1. Why is the Midwest good for growing corn?

   _____

   _____

2. Name three foods made with corn.

   _____

   _____

3. Why is growing corn important?

   _____

   _____

WEEK 25 DAY 4

Name: _____ **Date:** _____

Directions: Each piece of corn on an ear of corn is called a *kernel*. Study the table. Then, answer the questions.

Number of Kernels	Amount
on an average ear of corn	800
in one pound	1,300
in a bag of microwave popcorn	100

1. How many bags of microwave popcorn can be made from one ear of corn?

   _____

2. Does it take more or less than one ear of corn to get a pound of kernels? How do you know?

   _____

   _____

3. How many kernels would be in 4 average ears of corn? How did you find your answer?

   _____

   _____

   _____

Challenge: Draw pictures in the margins to support your answers.

WEEK 25 DAY 5

Name: _____ Date: _____

Directions: Corn is used to feed chickens and cows. If you ate chicken or a hamburger, thank corn. Corn is used in many foods, too. Make a list of all the things you ate today. Highlight the ones that use corn.

Geography and Me

WEEK 26 DAY 1

Name: _____ Date: _____

Directions: This map shows the states that collect the most solar energy. Solar energy comes from the sun. It can be turned into electricity. Study the map. Then, answer the questions.

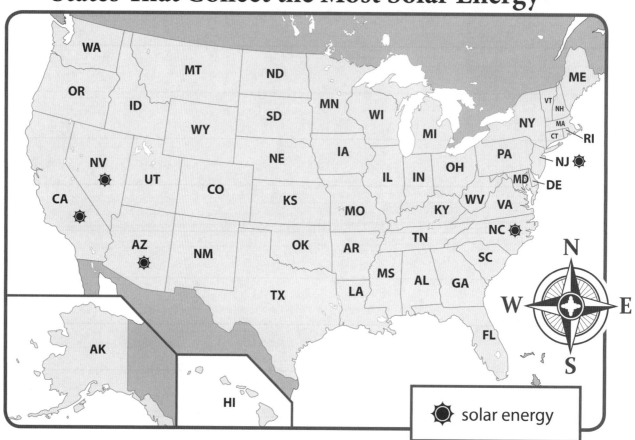

1. Circle the map features you see.

2. Do you think the map has a fitting title? Why or why not?

   _____

   _____

3. Which states collect the most solar energy?

   _____

   _____

WEEK 26 DAY 2

Name: _____ Date: _____

Directions: Below is a list of the top three states that use other types of energy. Place symbols on the map to show the types of energy used. Add your symbols to the legend. Then, give the map a new title.

Top states for coal: Wyoming (WY), West Virginia (WV), Kentucky (KY)

Top states for natural gas: Texas (TX), Pennsylvania (PA), Oklahoma (OK)

Top states for oil: Texas (TX), North Dakota (ND), Alaska (AK)

Creating Maps

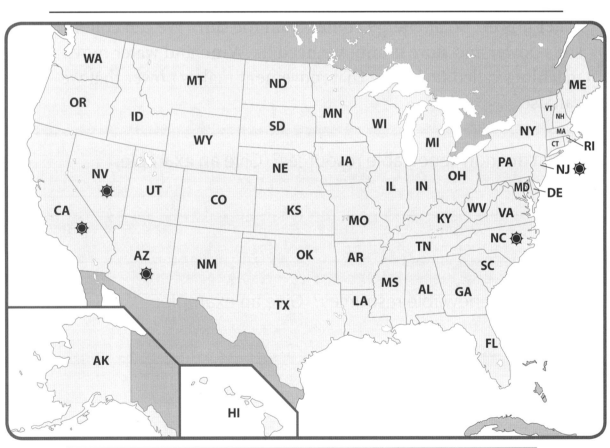

Legend

☀ solar energy

WEEK 26 DAY 3

Name: _____ **Date:** _____

Directions: Read the text. Then, answer the questions.

Sources of Power

We use electricity each day. We turn on lights. We watch TV. We use phones and computers. All of these things use electricity. That power is made in different ways.

Most places make power from non-renewable sources. When we use them, they are gone. We cannot make more of them. These include things such as coal, natural gas, and oil.

Some places make power from renewable sources. Nature can make these sources again. These are things such as solar, wind, and water power. Solar energy comes from the sun. We can collect the sun's power and store it until we need it. Wind and water move machines called *turbines*. That movement makes power that we can use.

1. What are non-renewable resources? Give an example.

   _____

   _____

2. What are renewable resources? Give an example.

   _____

   _____

3. Why might people want to use renewable resources?

   _____

   _____

Name: _____ Date: _____

Directions: The United States uses renewable and non-renewable resources to make electricity. Sort the resources into categories.

 water   wind

 coal   natural gas

 oil   solar

Renewable Resources	Non-Renewable Resources

WEEK 26 DAY 5

Name: _____ Date: _____

Directions: Draw a map of your classroom. Label everything that uses electricity.

Name: _____ Date: _____

Directions: Study the map. Then, answer the questions.

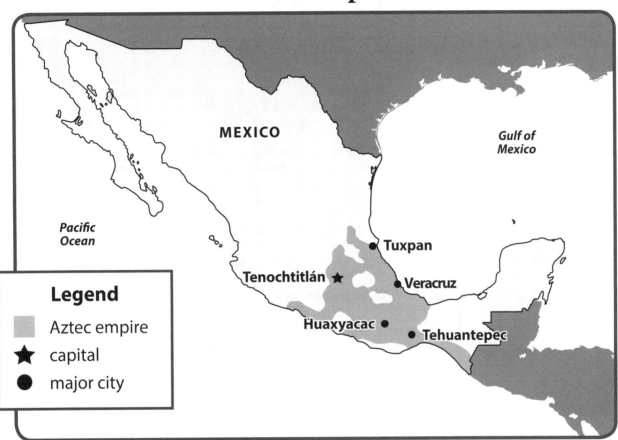

1. What is the capital of the Aztec Empire?

   _____

2. The Aztec capital was near a lake. Why might the Aztecs have built their capital near a body of water?

   _____

   _____

3. In what country was the Aztec Empire?

   _____

Name: _____ **Date:** _____

Directions: Follow the steps to complete the map.

Aztec Empire

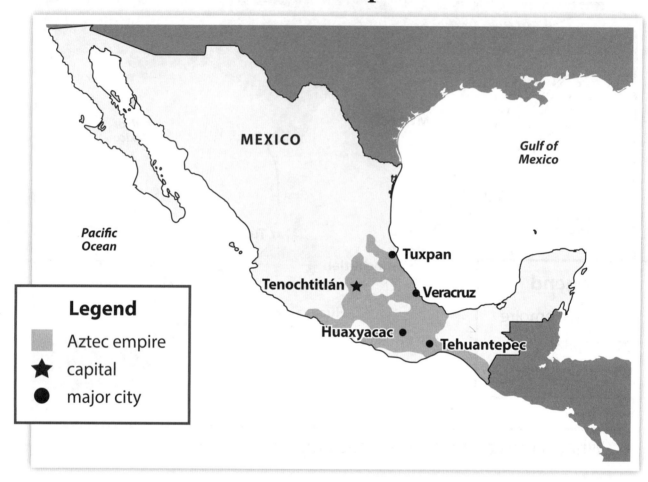

1. Hernán Cortés sailed from Spain to conquer Mexico. Draw his ship in the Gulf of Mexico.

2. Cortés landed on the east coast of the Aztec Empire near Veracruz. Draw a path from Veracruz to Tenochtitlán.

3. Outline the Aztec Empire in red.

Challenge: Look up a map of the Aztec Empire. Ask an adult to help you. Add cities and other features that you find to the map above.

Name: _____ **Date:** _____

Directions: Read the text. Then, answer the questions.

The Aztec Empire

The Aztecs ruled parts of Mexico for hundreds of years. They had their own culture and way of life. They made a calendar to keep track of holidays. The Aztecs built roads and temples. They farmed the land. They traded crops and other goods with other people in the region. They built an empire.

In 1519, Hernán Cortés invaded the Aztecs. He wanted to gain more land for Spain. The Spanish wanted gold and silver. They wanted to rule the region. The Spanish enslaved Aztec people. Many died from disease. This ended the Aztec Empire.

Yet, some of the Aztecs lived. They passed on their language. Some people still speak it. Today, people learn about Aztec culture.

1. Why did the Aztecs make a calendar?

   _____

   _____

2. Why did the Spanish invade the Aztecs?

   _____

   _____

3. How did the Spanish affect the Aztec Empire?

   _____

   _____

WEEK 27 DAY 4

Name: _____ **Date:** _____

Directions: The table shows dates of events in Mexican history. Study the table, and answer the questions.

Date	Event
1345	The Aztec city of Tenochtitlán is founded.
1428	The Aztecs conquer their rivals and expand their empire.
1517	Hernán Cortés comes to Mexico for the first time. He had 3 ships and 100 men.
1519	Cortés returns with 11 ships and 400 men. He conquers the Aztecs for Spain.
1821	Mexico wins its freedom from Spain.

1. When was the Aztec city of Tenochtitlán founded?

   _____

2. When did Cortés conquer the Aztecs?

   _____

3. How long did Spain control Mexico? How do you know?

   _____

   _____

4. How did Spain change Mexico?

   _____

   _____

   _____

WEEK 27 DAY 5

Name: _____ Date: _____

Directions: The Aztecs traded many things by bartering. Bartering is when you trade without using money. Make a list of items for a class party. Write things you could trade to get those items.

Items for a Party	Things We Could Trade

Geography and Me

WEEK 28 DAY 1

Name: _____ **Date:** _____

Directions: Study the map. Then, answer the questions.

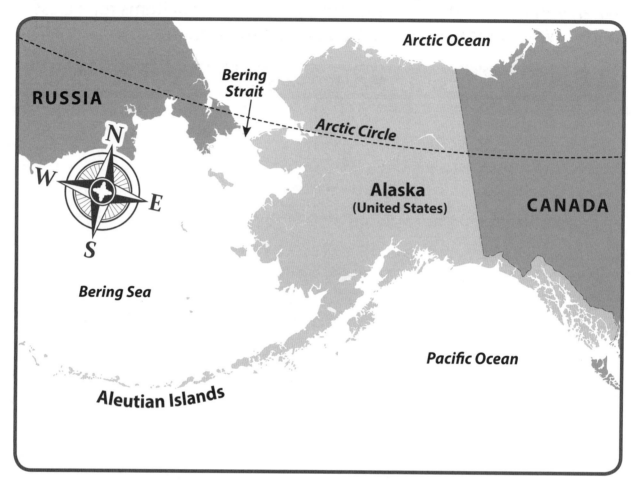

1. What countries border the Bering Strait?

   _____

2. What ocean is north of Alaska?

   _____

3. What ocean is south of Alaska?

   _____

Name: _____ Date: _____

Directions: Follow the steps to complete the map.

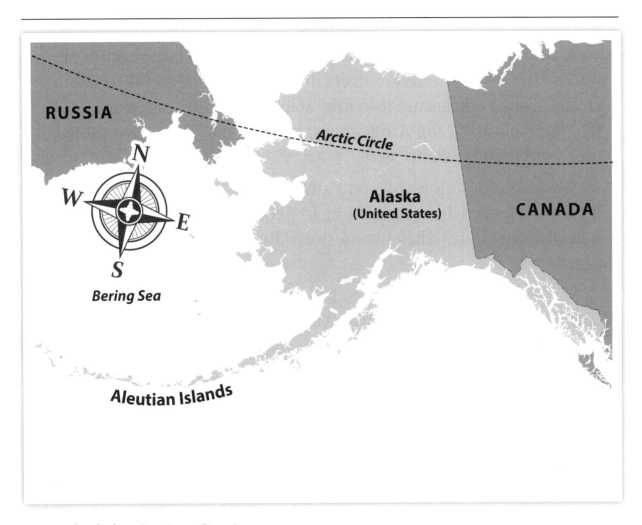

1. Label the Bering Strait.

2. Label the Arctic Ocean.

3. Label the Pacific Ocean.

4. Trace the border between Alaska and Canada.

5. Draw a path through the Bering Strait.

6. Add a title to the map.

WEEK 28 DAY 3

Name: _____ Date: _____

Directions: Read the text. Then, answer the questions.

The Bering Strait

A strait is a narrow waterway that connects two large bodies of water. The Bering Strait connects the Arctic Ocean and the Pacific Ocean. Some whales use it to migrate. They swim from their feeding grounds in the Arctic to warmer waters. They have babies there and then return to feed.

Some Alaska Natives live on a small island in the Bering Strait. It is called Little Diomede. About 70 people live there. They have a school and a clinic. They have a store. But there are no cars. Few people visit.

Scientists think there used to be land in the Bering Strait. It used to connect Asia and North America. Long ago, people walked across this land. Now, it is filled with water.

1. What is a strait?

_____

_____

2. Which oceans does the Bering Strait connect?

_____

_____

3. What is life like on Little Diomede?

_____

_____

WEEK 28 DAY 4

Name: _____ Date: _____

Directions: Every year, the sea ice in the Bering Strait is melting more. Study the graph, and answer the questions.

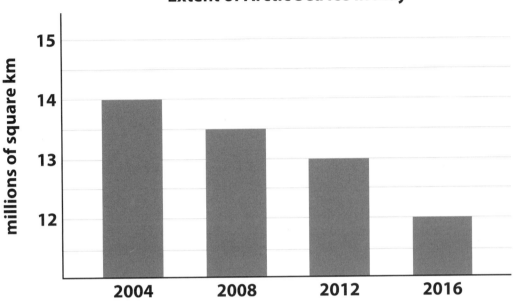

1. Which year had the most sea ice?

   _____

2. What do you notice about the data from 2004 to 2016?

   _____

   _____

   _____

3. What do you predict the data for 2020 might look like?

   _____

   _____

   _____

   _____

Think About It

**WEEK 28
DAY 5**

Name: _____ Date: _____

Directions: The United States makes rules to keep ships safe in the Bering Strait. Think about how your class travels down the hallway. Write and draw the rules that keep everyone safe in the hallways.

WEEK 29 DAY 1

Name: _____ Date: _____

Directions: Study the map. Then, answer the questions.

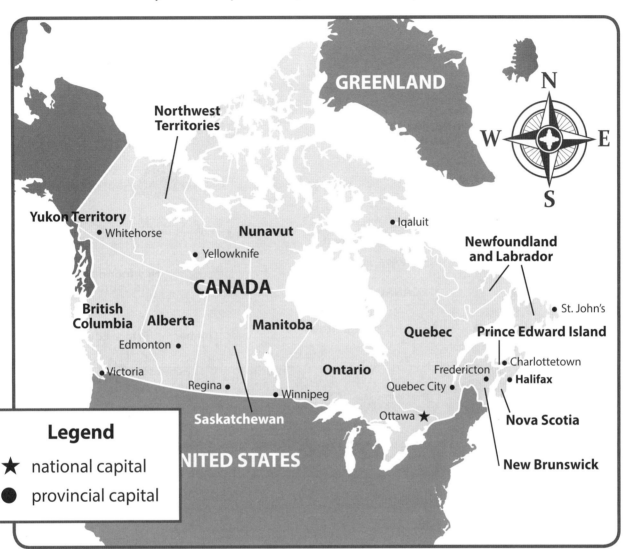

1. Canada is north of which country?

   _____

2. What is the capital of Canada?

   _____

3. What is the capital of Manitoba?

   _____

WEEK 29 DAY 2

Name: _____ Date: _____

Directions: Follow the steps to complete the map.

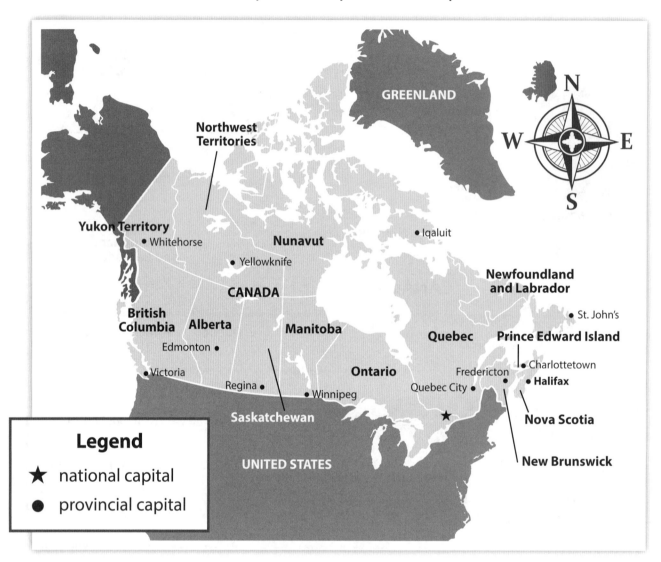

1. Label Ottawa, the capital of Canada.

2. Shade Alberta orange.

3. Shade Quebec blue.

4. Trace the borders between all provinces green.

5. Trace the borders with the United States red. Remember, Alaska is a U.S. state west of Canada.

WEEK 29 DAY 3

Name: _____ Date: _____

Directions: Read the text. Then, answer the questions.

Settlers in Canada

In 1534, Jacques Cartier explored the St. Lawrence River. He claimed the land for France. Later, a French settlement was started in Quebec. It was the first in North America. Over the years, the French and the British fought. In 1763, the French gave all their land in North America to the British. Today, both English and French are spoken there.

There are three native groups in Canada. The First Nations lost land to the settlers. Settlers also killed the bison they depended on. Many First Nations died of diseases the settlers brought. The Inuit lived in the Arctic. Contact with the settlers changed their way of life and their language. The third group is the Metis. Many of these groups still fight for their rights.

1. What does it mean to claim land for France?

   _____

   _____

2. Where was the first French settlement in North America?

   _____

3. What was the effect of the settlers on the native people?

   _____

   _____

   _____

WEEK 29 DAY 4

Name: _____ **Date:** _____

Directions: The table shows the official languages spoken in Canada. Study the table. Then, answer the questions.

Land Area	Province or Territory	Official Languages
Alberta	Province	English
British Columbia	Province	English
Manitoba	Province	English
New Brunswick	Province	English and French
Newfoundland and Labrador	Province	English
Nova Scotia	Province	English
Ontario	Province	English
Prince Edward Island	Province	English
Quebec	Province	French
Saskatchewan	Province	English
Yukon	Territory	English and French
Nunavut	Territory	English and French, Inuktitut, and Inuinnaqtun
Northwest Territories	Territory	English, French, and nine native languages

1. How many provinces and territories have French and English as official languages?

   _____

2. Think about Canada's history. Why might French and English be official languages?

   _____

   _____

   _____

158 28623—180 Days of Geography © Shell Education

WEEK 29 DAY 5

Name: _____ **Date:** _____

Directions: Canada and the United States have to solve problems together. Imagine your desk is a country. The desk next to you is a different country. Create a list of rules to help you get along.

Geography and Me

WEEK 30 DAY 1

Name: _____ Date: _____

Directions: Study the map. Then, answer the questions.

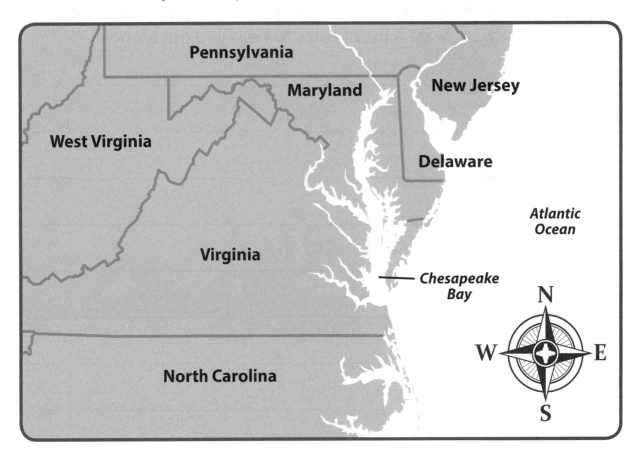

1. Which two states surround the Chesapeake Bay?

   _____

2. What ocean is connected to the Chesapeake Bay?

   _____

3. Is the Chesapeake Bay on the east or west coast of the United States? How do you know?

   _____

   _____

   _____

Name: _____ Date: _____

Directions: Follow the steps to complete the map.

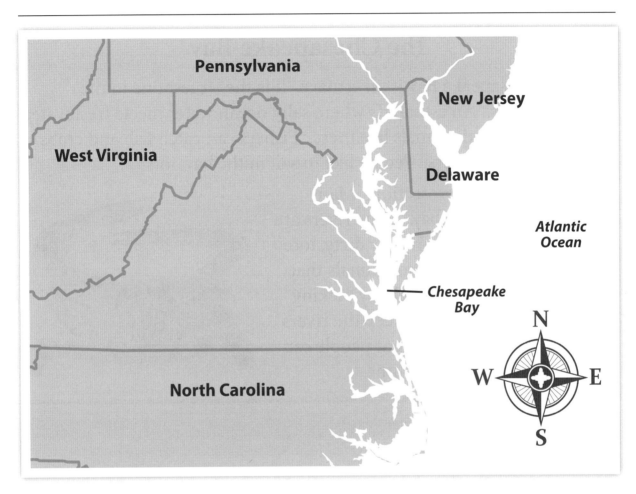

1. Label Virginia.

2. Label Maryland.

3. Label the Chesapeake Bay. Color it blue.

4. Title the map: The Chesapeake Bay.

Challenge: Cover the names of the states with sticky notes. Label as many states as you can. Then, remove the sticky notes to check your answers.

Name: _____ **Date:** _____

Directions: Read the text, and study the photo. Then, answer the questions.

The Chesapeake Bay

The Chesapeake Bay is an estuary. It is the largest one in the United States. An estuary is where salty ocean water meets freshwater. Many plants and animals live there. Fishers can catch fish and crabs. They look for oysters. People like to sail in the bay, too.

The bay needs protection. Many rivers flow into the bay. If the rivers are dirty, they can make the bay dirty, too. This hurts the plants and animals that live there. People can help by picking up trash. They can also keep the rivers clean. That will ensure that people can enjoy the bay for a long time.

1. What is an estuary?

   _____

2. What do people like to do in Chesapeake Bay?

   _____

   _____

3. What can people do to protect the bay?

   _____

   _____

Name: _____ Date: _____

Directions: A watershed is land that drains into a body of water. Everything that happens in the Chesapeake Bay watershed affects the bay. Study the table and answer the questions.

The Chesapeake Bay Watershed	
people living in the watershed	18 million
land in the watershed	64,000 square miles (166,000 square km)
states in the watershed	Delaware, Maryland, New York, Pennsylvania, Virginia, West Virginia

1. How many people live in the Chesapeake Bay watershed?

   _____

2. How much land is in the Chesapeake Bay watershed?

   _____

3. How many states are in the watershed? How might this make it difficult to protect the bay?

   _____

   _____

   _____

4. Why might it be important to keep the watershed clean?

   _____

   _____

WEEK 30 DAY 5

Name: _____ Date: _____

Directions: One way to protect our lands and water is to upcycle. This is where you take something that would be thrown out and turn it into art. Draw and write to show how you could upcycle something.

Geography and Me

Name: _____ Date: _____

Directions: This is a map of the Colorado River. Study the map, and answer the questions.

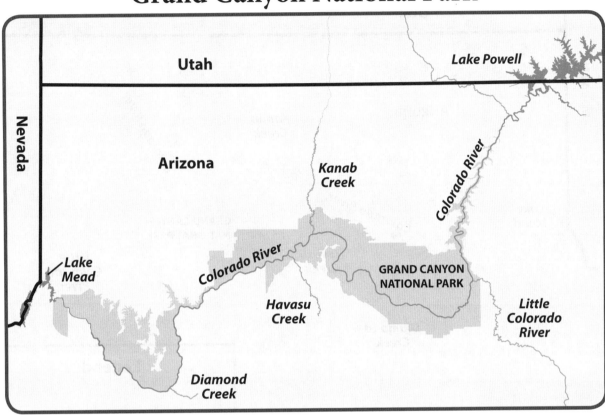

1. What state is the Colorado River in?

   _____

2. What national park does the Colorado River flow through?

   _____

3. What rivers flow into the Colorado River?

   _____

   _____

   _____

WEEK 31 DAY 2

Name: _____ **Date:** _____

Directions: Follow the steps to complete the map.

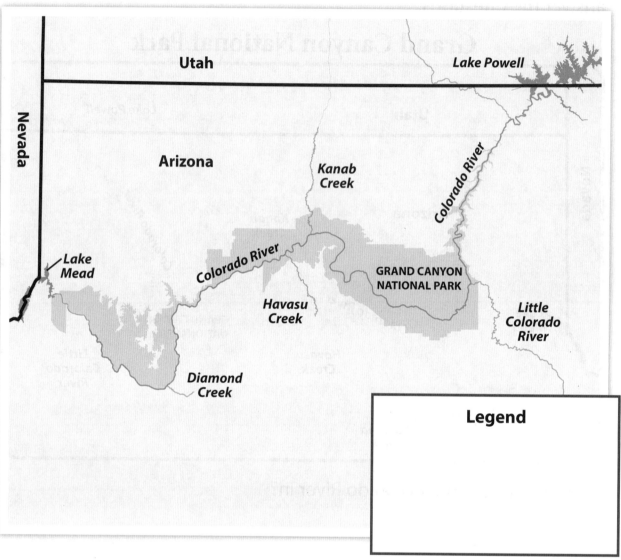

1. Trace the Colorado River in blue.

2. Outline Grand Canyon National Park in orange.

3. Complete the legend to show what the colors mean.

Name: _____ **Date:** _____

Directions: Read the text, and study the photo. Then, answer the questions.

The Colorado River

The Grand Canyon has a river at the bottom of it. It has steep sides. It is a mile deep in some places. Long ago, the Grand Canyon did not exist. Over time, the Colorado River eroded the rocks. That means it washed the rocks away. The canyon got deeper and deeper. The river is still digging slowly through the rocks.

People today take water from the Colorado River. They use it for drinking and farming. People take more water from this river than from any other. The river has many dams. A dam controls the flow of the river. The dams help get water to people. Some dams use the river to make power.

1. How did the Colorado River make the Grand Canyon?

   _____

2. How deep is the Grand Canyon?

   _____

3. What does a dam do?

   _____

   _____

WEEK 31 DAY 4

Name: _____   **Date:** _____

Directions: Lake Mead was made when people built a dam on the Colorado River. Three states get water from Lake Mead. Study the photo of Lake Mead, and answer the questions.

1. Trace the top edge of the white band above the water. This is how high the water used to be.

2. Is there more or less water in Lake Mead now? How do you know?

   _____

   _____

3. What might that mean for the states that take water from Lake Mead?

   _____

   _____

WEEK 31 DAY 5

Name: _____ **Date:** _____

Directions: Draw a comic strip showing all the ways you used water today.

Geography and Me

WEEK 32
DAY 1

Name: _____ Date: _____

Directions: Study the map. Then, answer the questions.

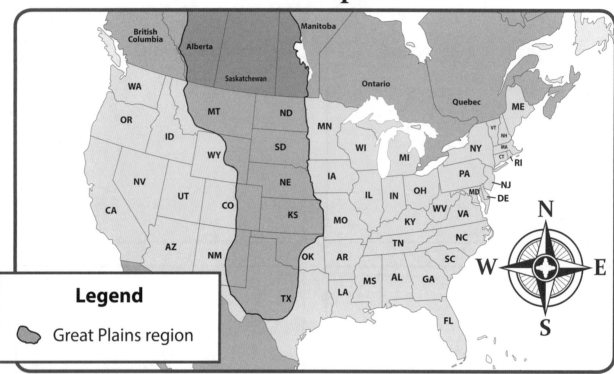

1. Which U.S. states are in the Great Plains region?

   _____

   _____

2. Which Canadian provinces are in the Great Plains region?

   _____

   _____

3. Which states are entirely in the Great Plains region?

   _____

   _____

Name: _____ Date: _____

Directions: Bison lived on the Great Plains. They moved north toward Canada in the summer. They came back south in the winter. Follow the steps to complete the map.

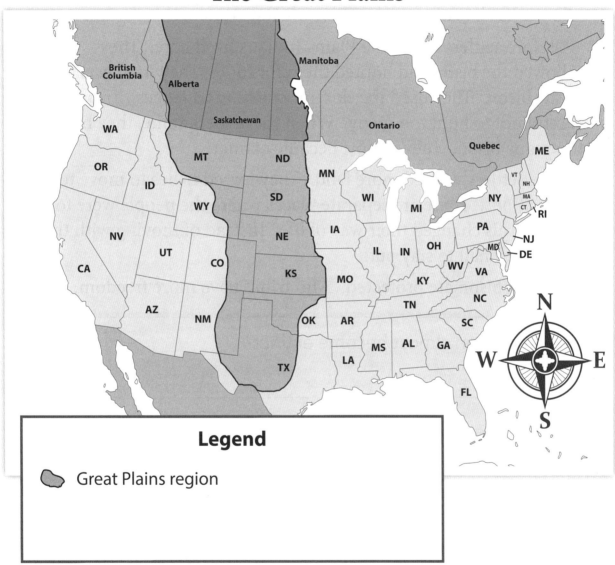

1. Beginning in the middle of the Great Plains, draw a path the bison might take north.

2. Use another color to draw the return path south.

3. Complete the legend to show what the colors mean.

WEEK 32 DAY 3

Name: _____ Date: _____

Directions: Read the text. Then, answer the questions.

The Great Plains

The Great Plains is a region in North America. It has rolling hills. It has lots of grasses. There are few trees in this region.

Before settlers came, the Plains Indians lived here. They followed the bison and hunted them for food. They used every part of the bison. They used the skin for clothes and tepees. Tribes were careful not to hunt too many. When the settlers came, they hunted too many bison. They almost became extinct.

The settlers wanted more land. They forced tribes to move to reservations. These are separate places where the people were told to live. This changed their way of life. Instead of moving with the bison, they had to stay put.

Today, bison are protected. And tribes have more freedom.

1. How did bison almost become extinct?

   _____

2. What effect did the settlers have on the Plains Indians?

   _____

   _____

3. What was the result of the Plains Indians having to move to reservations?

   _____

   _____

WEEK 32 DAY 4

Name: _____ Date: _____

Directions: Plains Indians moved from place to place to follow herds of bison. They lived in tepees. Teepees could be taken down and moved quickly. Study the picture. Then, answer the questions.

1. Describe what a tepee looks like.

   _____

   _____

2. Why did a tepee make a good home for the Plains Indians?

   _____

   _____

3. How is a tepee like a tent that is used today for camping? How is it different?

   _____

   _____

   _____

WEEK 32 DAY 5

Geography and Me

Name: _____ Date: _____

Directions: Imagine you are visiting the Great Plains. Write a journal entry to describe what you see.

Name: _____ Date: _____

Directions: This map shows the Transcontinental Railroad. One company started building in Sacramento. Another started in Omaha. They met in Utah in 1869. Study the map, and answer the questions.

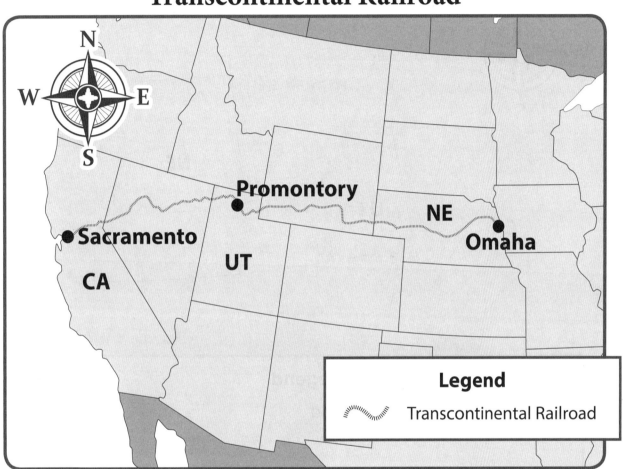

1. Where did the Transcontinental Railroad start in the East?

   _____

2. Where did the Transcontinental Railroad start in the West?

   _____

3. Where did the two companies meet?

   _____

Name: _____ **Date:** _____

Directions: Follow the steps to complete the map.

Transcontinental Railroad

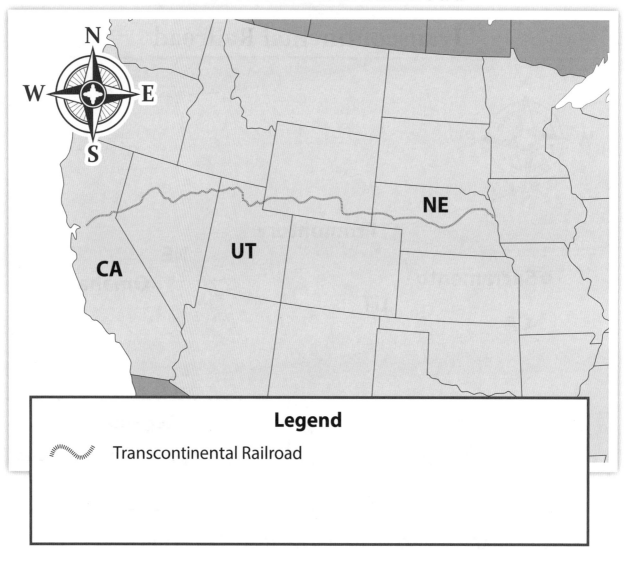

1. Add a symbol to the map to show where the Transcontinental Railroad began in the East.

2. Add a symbol to show where it began in the West.

3. Add a symbol to show where the companies met.

4. Complete the legend to show what the symbols mean.

Name: _____ **Date:** _____

Directions: Read the text, and study the photo. Then, answer the questions.

A New Railroad

Transcontinental means to go across the continent. In the 1860s, two companies started building one long railroad. One group started in the East. Another started in the West. They met in Utah in 1869. They drove a Golden Spike to hold the last piece of track.

The trains moved people west. They also moved mail and supplies. Trains made travel faster and safer. Before, it took six months to travel west. Now, it took one week. More people moved west to live and work.

1. How did the Transcontinental Railroad change travel to the West?

   _____

   _____

   _____

2. What effect did the Transcontinental Railroad have on the West?

   _____

   _____

   _____

WEEK 33 DAY 4

Name: _____ **Date:** _____

Directions: The table shows how long it took to travel from the East to the West in the 1800s. Study the table. Then, answer the questions.

Transportation	Time
ship	5–7 months
horse and wagon	3–6 months
train	1 week

1. What form of transportation was the shortest?

   _____

2. What form of transportation was the longest?

   _____

3. Imagine you were traveling from the East to the West in the 1800s. How would you want to travel? Why?

   _____

   _____

   _____

   _____

Challenge: Look up how long it would take to travel from Omaha, Nebraska, to Sacramento, California, today. Ask an adult to help you. See how long it would take to travel by car and by plane.

178   28623—180 Days of Geography   © Shell Education

WEEK 33 DAY 5

Name: _____ **Date:** _____

Directions: What do you think will be the next form of transportation? Write and draw to show how you think people will get around in the future.

Geography and Me

WEEK 34 DAY 1

Name: _____   Date: _____

Directions: This map shows the four main regions of the United States. Study the map. Then, answer the questions.

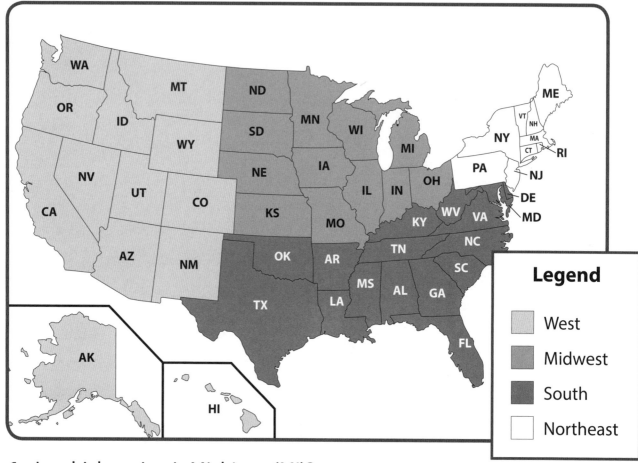

1. In which region is Michigan (MI)?

   _____

2. In which region is Texas (TX)?

   _____

3. In which region is Pennsylvania (PA)?

   _____

4. Name three states in the western region.

   _____

Name: _____ **Date:** _____

Directions: Imagine you are in charge of creating regions of the United States. Show how you would divide the country on the map. Then, complete the legend to show your regions.

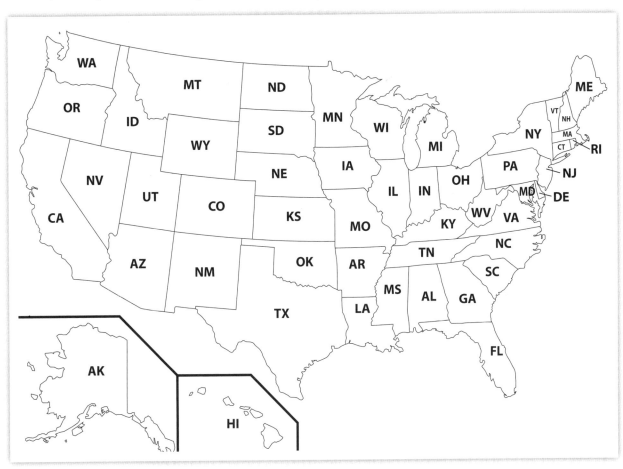

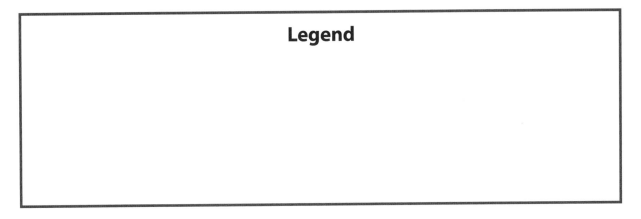

Legend

WEEK 34 DAY 3

Name: _____ Date: _____

Directions: Read the text. Then, answer the questions.

Population

Population is the number of people in a place. It can change over time. People move for many reasons. They may move for better weather. People may want to be near family. People move for their jobs.

In the Northeast and the Midwest, many factories have closed. Some people there cannot find new jobs. They may have to move. Many people are moving to the South and the West. Homes cost less money. The weather is warmer. And there may be more jobs.

When the population changes, other things change. A place that is getting smaller might have too many teachers. There may be empty homes. Businesses may close. A growing place may need more schools. It may need more homes and roads. Growing places need more resources. People need power and water. Cities have to plan to get these new resources.

1. What are some reasons people might have to move?

   _____

   _____

2. Which regions are people moving from?

   _____

3. How does a growing population change a place?

   _____

   _____

Name: _____ **Date:** _____

Directions: This table shows how the population has grown in each region. Study the table. Then, answer the questions.

Region	Population in 2010	Population in 2016
Northeast	55 million	56 million
Midwest	67 million	68 million
South	115 million	122 million
West	72 million	76 million

1. How much did each region grow?

   _____

   _____

   _____

2. Which two regions grew the most?

   _____

   _____

3. Which two regions grew the least?

   _____

   _____

**WEEK 34
DAY 5**

Name: _____ Date: _____

Directions: Population is how many people are in a place. Think about the populations at your school. Then, answer the questions.

1. What is the population of your class?

   _____

2. How many other second grade classes are in your school?

   _____

3. What is the population of second grade at your school?

   _____

4. Draw a picture to show each of these populations.

Name: _____ Date: _____

Directions: Study the map. Then, answer the questions.

Yellowstone and Grand Teton National Parks

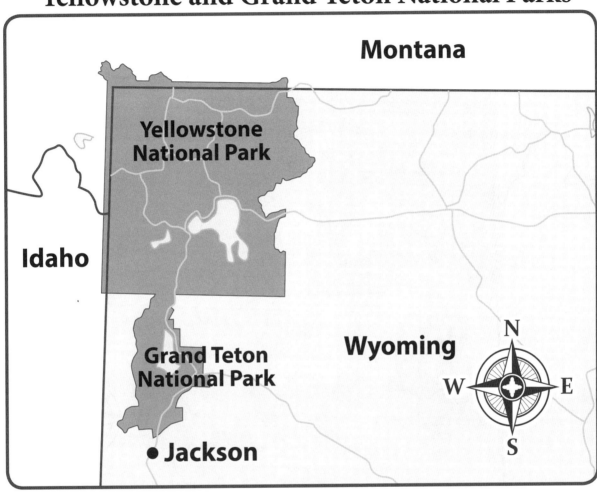

1. In which three states is Yellowstone National Park?

   _____

   _____

2. What direction is Jackson from Yellowstone?

   _____

3. What national park is south of Yellowstone?

   _____

Name: _____ Date: _____

Directions: Follow the steps to complete the map.

Yellowstone and Grand Teton National Parks

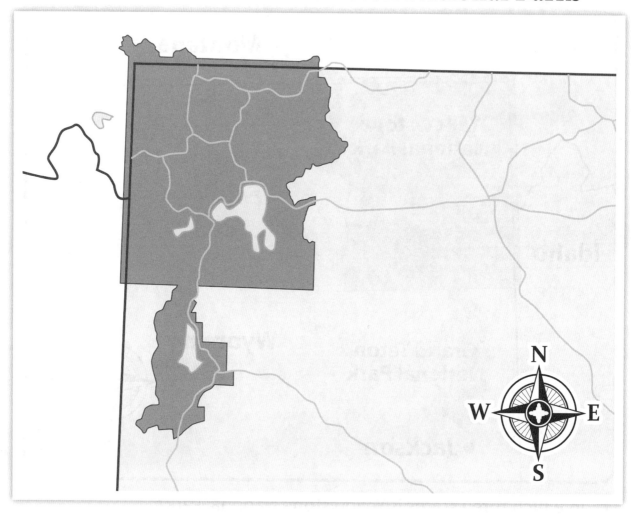

1. Label Wyoming, Idaho, and Montana. Most of the park is in Wyoming. Idaho is to the west. Montana is to the north.

2. Outline Yellowstone National Park in green.

3. Outline Grand Teton National Park in blue. It is south of Yellowstone.

4. Draw a route from the west side of Yellowstone to the east side. Be sure your route stays on the roads.

Bringing Back the Wolves

When settlers moved west, they saw wolves as a threat. Wolves sometimes killed cows and sheep. So settlers killed the wolves. In 1926, the last group of wolves in Yellowstone was killed. People knew that the wolves were a key part of the ecosystem. That is all the living and non-living things in a place. In 1974, the wolf became an endangered species.

People worked to bring wolves back to Yellowstone. It is a safe place for the wolves. Now, there are more than 100 wolves in the park. There are more than 500 in the area around it. The wolves eat mostly elk. People worry that the elk may be endangered next.

1. When was the last group of wolves killed in Yellowstone?

   _____

2. When were the wolves named an endangered species?

   _____

3. What is the effect of more wolves in Yellowstone National Park?

   _____

   _____

   _____

WEEK 35
DAY 4

Name: _____ Date: _____

Directions: The table shows the numbers of species in Yellowstone National Park. Study the table. Then, answer the questions.

Type of Animals	Number of species
mammals	67
birds	300
fish	16
amphibians	5
reptiles	6

1. Which type of animal has the most species in Yellowstone?

   _____

2. Which type of animal has the least amount of species in Yellowstone?

   _____

3. How many more mammal species are there than reptiles?

   _____

   _____

4. Based on the chart, how would you describe Yellowstone?

   _____

   _____

   _____

WEEK 35 DAY 5

Name: _____ Date: _____

Directions: The third Friday in May is Endangered Species Day. It is a day for people to learn about protecting endangered animals. Design a poster that tells people about Endangered Species Day.

Geography and Me

WEEK 36 DAY 1

Name: _____ Date: _____

Directions: Study the map and answer the questions.

Top Producers of Fruits and Vegetables

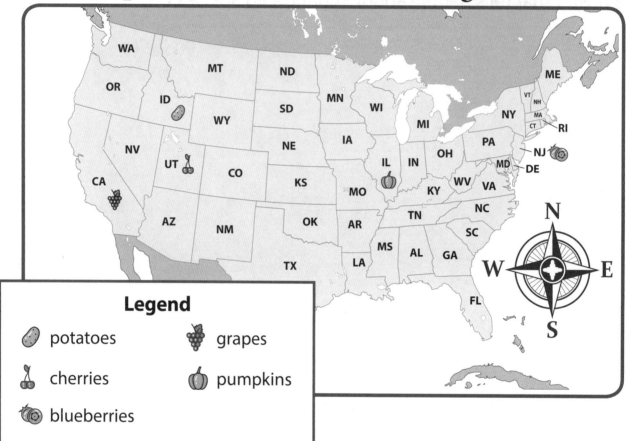

1. Which state grows the most pumpkins?

   _____

2. Which state grows the most grapes?

   _____

3. Which state grows the most potatoes?

   _____

4. Which state grows the most cherries?

   _____

Name: _____ **Date:** _____

Directions: The states listed in the box grow a lot of apples. Draw a symbol on each of those states. Then, add your symbol to the legend.

Connecticut (CT)	New York (NY)	Rhode Island (RI)
Minnesota (MN)	Ohio (OH)	Washington (WA)
New Hampshire (NH)	Pennsylvania (PA)	West Virginia (WV)

Top Producers of Fruits and Vegetables

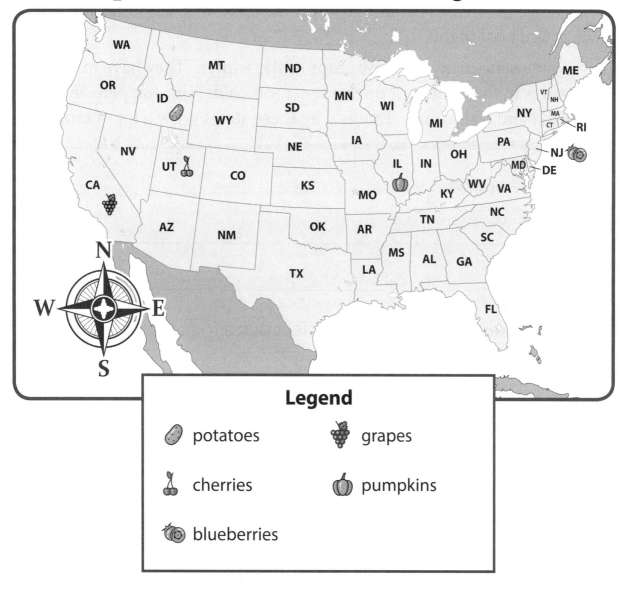

WEEK 36 DAY 2

Creating Maps

WEEK 36 DAY 3

Name: _____ Date: _____

Directions: Read the text. Then, answer the questions.

Crops in the United States

Climate is what the weather is usually like. The climate is different in different places. Plants have adapted to grow well in certain climates. This means crops grow best in certain places. They may not grow well in places with different climates.

Washington grows the most apples. It has a dry, warm climate. This means diseases cannot grow on the apples. It has long, sunny days and cool nights.

Some states can still grow crops in the winter. They have mild climates in the winter. It does not get too cold. This happens in Florida and California. That is why these states grow a lot of crops.

1. What is climate?

   _____

   _____

2. What makes Washington the perfect place to grow apples?

   _____

   _____

3. Why can Florida and California grow crops in the winter?

   _____

   _____

Name: _____ **Date:** _____

Directions: The table shows the average high temperatures in different regions. The high temperature is the warmest it gets during the day. Study the table. Then, answer the questions.

U.S. Region	High Temperature in January	High Temperature in June
Northeast	30°F (-1°C)	70°F (21°C)
Midwest	20°F (-6°C)	80°F (26°C)
South	60°F (15°C)	90°F (32°C)
West	60°F (15°C)	90°F (32°C)

1. Which region is the coldest in January?

   _____

2. Which regions have the highest temperatures in June?

   _____

3. Which region changes the most between January and June? How many degrees Fahrenheit does it change?

   _____

   _____

4. In which region would you most want to live? Why?

   _____

   _____

   _____

WEEK 36
DAY 5

Name: _____ Date: _____

Directions: Think about where you live. Draw what you would wear outside in January. Then, draw what you would wear outside in June. Under each box, write a sentence about the weather that month.

January	June

_____   _____

_____   _____

_____   _____

_____   _____

_____   _____

ANSWER KEY

There are many open-ended pages and writing prompts in this book. For those activities, the answers will vary. Answers are only given in this answer key if they are specific.

Week 1 Day 1 (page 15)
1. title
2. compass rose
3. legend

Week 1 Day 2 (page 16)
1. triangles
2. trees
3. a dot

Week 1 Day 3 (page 17)
1. Gold River
2. east
3. Crystal Mountains

Week 2 Day 2 (page 21)
1. Atlantic Ocean
2. Pacific Ocean
3. Arctic Ocean

Week 2 Day 3 (page 22)
1. Canada
2. Mexico
3. Canada

Week 3 Day 1 (page 25)
1. north
2. west
3. south and west

Week 3 Day 3 (page 27)
1. Answers may include that it can help people get around town, such as a bus driver.
2. A globe is shaped like Earth, while a map is flat.

Week 3 Day 4 (page 28)
1. Answers may include that they show the same place.
2. Answers may include that they show different points of view.
3. Students should include details from the pictures in their answers.

Week 4 Day 1 (page 30)
1. corn and soybeans
2. fruits and vegetables
3. meat

Week 4 Day 3 (page 32)
1. Transportation is the movement of people and things.
2. Companies choose the transportation that will save them money.
3. Answers may include that people choose the right way for where they want to go.

Week 4 Day 4 (page 33)
1. Answers may include that the trip is long because it takes multiple days.
2. airplane
3. Answers should be supported with details from the chart.

Week 5 Day 1 (page 35)

Land should be colored green, and water should be colored blue. There is more water.

Week 5 Day 2 (page 36)

left: Pacific Ocean
top: Arctic Ocean,
right: Atlantic Ocean

Week 5 Day 3 (page 37)
1. We can only use a small part of the water.
2. Less trash ends up in the water, and it makes the water cleaner.
3. Answers may include picking up trash, recycling, and not wasting water.

Week 5 Day 4 (page 38)
1. 15 gallons (57 L)
2. 4 gallons (15 L)
3. filling the tub halfway; it saves more gallons than the other ways.
4. Answers may include watering plants less often, turning the water off while washing a car, etc.

Week 6 Day 2 (page 41)

From top to bottom: Canada, United States, Mexico

Week 6 Day 3 (page 42)
1. so leaders can make laws for their people
2. a river called the Rio Grande
3. Answers may include that people can change borders to solve problems or after a war.

© Shell Education

ANSWER KEY (cont.)

Week 6 Day 4 (page 43)
1. Canada and the United States
2. Spanish
3. Answers may include that communication is more difficult when countries speak different languages.
4. Answer should be supported by reasons from the text or personal experience.

Week 7 Day 1 (page 45)
1. rural; it is on a farm, and there is a lot of land around it.
2. suburban
3. Answers may include that the rural land will shrink.

Week 7 Day 2 (page 46)
The top half is rural and the bottom is a city.

Week 7 Day 3 (page 47)
1. urban
2. rural
3. rural and urban; one is very crowded and the other is very spread out.

Week 7 Day 4 (page 48)
1. Nevada and Florida
2. Michigan; the chart shows a smaller number of people in 2010.
3. smaller; the number of people in the state decreased.

Week 8 Day 1 (page 50)
1. Nevada, Utah, Idaho, Oregon, California
2. Sierra Nevada Mountains
3. Nevada

Week 8 Day 3 (page 52)
1. It seeps into the ground.
2. Answers may include that the climate is different.

Week 8 Day 4 (page 53)
1. Badwater Basin
2. Mount Whitney
3. 10,611 ft. (3,234 m.)
4. Mount Whitney; it is the highest elevation.

Week 9 Day 1 (page 55)
1. Wisconsin, Ohio, Missouri
2. Answers may include that they were largely farmers.
3. Answers may include that it was a long and difficult trip.

Week 9 Day 3 (page 57)
1. putting up Christmas trees
2. There would be no kindergarten or PE classes.
3. noodle, hamburger

Week 9 Day 4 (page 58)
1. 6 horses
2. Answers may include on one of the horses or on the side of the wagon.
3. Answers may include that it helped prevent goods from falling out of the wagon.

Week 10 Day 1 (page 60)
1. Atlantic Ocean
2. Gulf of Mexico
3. Answers may include Texas, Oklahoma, Louisiana, Arkansas, Missouri, Illinois, Kentucky, Tennessee, Mississippi, Alabama, Georgia, Florida, South Carolina, Pennsylvania, North Carolina, Virginia, Maryland, Delaware, New Jersey, New York, Connecticut, Rhode Island, or Massachusetts.

Week 10 Day 3 (page 62)
1. a flat area of land next to an ocean
2. more than 20 states
3. Answers may include that they help transport resources.

Week 10 Day 4 (page 63)
1. a place where ships pick up or drop off goods to sell
2. large containers are loaded by cranes
3. Answers may include to transport their products by boat.

Week 11 Day 1 (page 65)
1. New Mexico
2. Cheyenne
3. Santa Fe
4. Santa Fe is near the center of the state. Cheyenne is near the southern border.

Week 11 Day 2 (page 66)
1. 17 people should be drawn.
2. Six people should be drawn.
3. New Mexico; it has more people per square mile.

Week 11 Day 3 (page 67)
1. The population goes up.
2. Answers may include to be closer to work, to be closer to family, to have more space, or for the weather.

ANSWER KEY (cont.)

Week 11 Day 4 (page 68)

1. New Mexico
2. Ohio
3. Answers should be supported with details from the table.
4. Answers should include that Ohio is more crowded than New Mexico or Wyoming.

Week 12 Day 1 (page 70)

1. New Orleans
2. Gulf of Mexico
3. south

Week 12 Day 3 (page 72)

1. 2,350 miles (3,800 km)
2. Answers may include that people use it for drinking, growing crops, and for transportation; animals live by the river and drink its water.
3. Answers may include that people can't eat the fish and it is bad for wildlife.

Week 12 Day 4 (page 73)

1. Answers may include that the city is larger, with more tall buildings.
2. Answers may include that the port is larger and has larger ships.
3. Answers may include that the ships are larger and are no longer steam-powered.

Week 13 Day 1 (page 75)

1. Lake Superior
2. Path should go through Lake Superior and Lake Huron, and end in Toronto.
3. four

Week 13 Day 3 (page 77)

1. five
2. The St. Lawrence Seaway connects the Great Lakes to the Atlantic Ocean.
3. Answers may include that the lakes give water to 48 million people, and they are home to fish and other wildlife.

Week 13 Day 4 (page 78)

1. Lake Erie
2. Lake Superior
3. 94 ft. (29 m); subtract the depth of Lake Michigan from the depth of Lake Huron.
4. Yes; it is the deepest lake.

Week 14 Day 1 (page 80)

1. north and east
2. north and west
3. Oak St.
4. the library

Week 14 Day 2 (page 81)

1. (D, 3)
2. Student should draw a house at (D, 6).
3. Student should draw a store at (A, 6).
4. Student should draw a fire station at (E, 3).
5. Student should draw a baseball field at (C, 1).

Week 14 Day 3 (page 82)

1. mail carrier
2. Answers may include bus drivers and crossing guards.
3. Answers may include fire fighters, police officers, doctors, and vets.
4. Answers may include raking leaves, walking dogs, or helping with other chores.

Week 14 Day 4 (page 83)

1. riding a bus
2. riding a bike
3. six students; answers should include subtracting the two numbers.
4. Answers should include details from the table or personal experience.

Week 15 Day 1 (page 85)

1. Missouri River
2. Hudson River
3. No, ocean is not freshwater. It is salty.

Week 15 Day 3 (page 87)

1. Answers may include cleaning up streams, planting plants, recycling, picking up litter, and taking shorter showers.
2. Only a small part of it is freshwater, and some of it is frozen.
3. Answers may include growing food, bathing, and drinking.

Week 15 Day 4 (page 88)

1. Answers may include plastic bottles, a soccer ball, plastic bags, and a box.
2. Answers may include not littering and picking up trash.
3. Answers may include littering or not recycling.

© Shell Education

ANSWER KEY (cont.)

Week 16 Day 1 (page 90)
1. three
2. east
3. west

Week 16 Day 3 (page 92)
1. Answers may include that people can work after dark, transportation and communication is easier, and air conditioning keeps us comfortable and cars let us live farther from work.
2. Answers may include that new homes, schools, roads, and towns can be built, and communities can grow.
3. Answers may include wood, gas, and water.

Week 16 Day 4 (page 93)
1. rural
2. urban
3. Answers may include that there are paved roads, cars, and many buildings.

Week 17 Day 1 (page 95)
1. Atlantic Ocean
2. Pacific Ocean
3. North America and South America

Week 17 Day 3 (page 97)
1. 50 miles (80 km)
2. Workers had to cut through mountains, the weather was bad, and many workers got sick.
3. Answers may include that it made travel faster and cheaper.

Week 17 Day 4 (page 98)
1. 14,000 miles (22,500 km)
2. 4,800 miles (7,700 km)
3. 9,200 miles (14,800 km)

Week 18 Day 1 (page 100)
1. the United States and Canada
2. New York
3. Ontario

Week 18 Day 3 (page 102)
1. They used candles.
2. Answers may include that cities grew, and more people moved to cities.
3. Answers may include that people could use safer lights, work later, shop later, and move to cities.

Week 18 Day 4 (page 103)
1. Answers may include that there are taller buildings and cars on the road.
2. Answers may include that the buildings are taller.
3. Answers may include that electricity helped the city grow.

Week 19 Day 1 (page 105)
1. United States, Mexico, Cuba
2. Alabama, Florida, Louisiana, Mississippi, and Texas
3. east

Week 19 Day 3 (page 107)
1. part of the ocean that is almost enclosed by land
2. Answers may include that people use the gulf for resources and transportation.
3. Answers may include that people and animals can get sick.

Week 19 Day 4 (page 108)
1. three
2. two
3. Answers may include that since most of these animals are endangered, they are in great danger.
4. It would become extinct.

Week 20 Day 1 (page 110)
1. The Thirteen Colonies
2. The compass rose should be circled.
3. colonies, American Indian territory, colony boundaries, French territory

Week 20 Day 3 (page 112)
1. They traded fur, tools, and food.
2. Answers may include that the settlers took over the land.
3. Answers may include that many American Indians died, they lost food sources, and they got diseases.

Week 20 Day 4 (page 113)
1. 20–30 million bison
2. 500,000 bison
3. Answers may include that it made a difference because the number of bison went up.
4. Answers should indicate that the number will stay about the same or increase.

Week 21 Day 1 (page 115)
1. four
2. Northeast, South, Midwest, and West
3. They tell where the regions are located.

ANSWER KEY (cont.)

Week 21 Day 3 (page 117)
1. 1902
2. Answers may include that more people moved to warmer regions.
3. Answers may include that people can stay inside when it's hot.

Week 21 Day 4 (page 118)
1. three
2. seven
3. Answers may include that cities in the South and the West grew after air conditioning became common.

Week 22 Day 1 (page 120)
1. Canada, the United States, and Mexico
2. Pacific Ocean
3. north and south; answers may include using the compass rose.

Week 22 Day 3 (page 122)
1. It splits the flow of rivers on the continent.
2. South Pass
3. Answers may include that they had a hard time crossing the high mountains in their wagons.

Week 22 Day 4 (page 123)
1. 96 pounds (43 kg)
2. 200 pounds (89 kg)
3. flour
4. Answers may include that hundreds of pounds of food are needed.

Week 23 Day 1 (page 125)
1. Canada
2. Aleutian Islands
3. Bering Strait

Week 23 Day 3 (page 127)
1. Answers may include that they hunted, fished, and used the land wisely.
2. The Russians brought diseases.
3. They use modern things, but they value their old ways of life.

Week 23 Day 4 (page 128)
1. baidarka; they would be traveling long distances over the water
2. umiak
3. sleds pulled by dogs
4. Example: I think sleds pulled by dogs was the fastest because dogs can run faster than people.

Week 24 Day 1 (page 130)
1. east
2. Lake Erie and Lake Ontario
3. Canada

Week 24 Day 3 (page 132)
1. Answers may include that the Erie Canal made it faster and easier for people to move west and to trade.
2. Answers may include that people thought the canal was a great achievement.

Week 24 Day 4 (page 133)
1. It made the trip much faster.
2. nine days
3. Answers may include that people wanted to get to their destinations faster or to move goods faster.
4. Answers may include that it made travel faster and easier.

Week 25 Day 1 (page 135)
1. four
2. 12 states
3. All 12 states should be colored orange.
4. The West, Midwest and South should all be colored different colors.

Week 25 Day 3 (page 137)
1. It has good soil, warm nights, and hot days.
2. Answers may include chewing gum, ketchup, taco shells, cereal, snack foods, and popcorn.
3. Answers may include that many products are made with corn.

Week 25 Day 4 (page 138)
1. 8 bags
2. more; one ear of corn has 800 kernels, and there are 1,300 in one pound.
3. 3,200 kernels; answers should include adding or multiplying the number of kernels on an ear of corn.

Week 26 Day 1 (page 140)
1. The map title, legend, and compass rose should be circled.
2. Answers may include that it is a fitting title because it describes what is shown on the map.
3. California, Arizona, Nevada, North Carolina, and New Jersey

ANSWER KEY (cont.)

Week 26 Day 3 (page 142)
1. Non-renewable resources are resources that, when we use them, are gone. Examples include coal, oil, and natural gas.
2. Renewable resources can be made again. Examples include solar, wind, and water.
3. Answers may include that renewable sources will not run out.

Week 26 Day 4 (page 143)
Renewable resources: solar, water, wind
Non-renewable resources: coal, oil, natural gas

Week 27 Day 1 (page 145)
1. The capital of the Aztec Empire is Tenochtitlán.
2. Answers may include that people need water to drink, cook, bathe, and for transportation.
3. Mexico

Week 27 Dway 3 (page 147)
1. to keep track of holidays
2. They wanted more land, gold, and silver.
3. Answers may include that they ended the Aztec Empire by spreading disease and enslaving the people.

Week 27 Day 4 (page 148)
1. 1345
2. 1519
3. 302 years; answers should include subtracting the years in the chart.
4. Answers should include that they took over the land or conquered the people.

Week 28 Day 1 (page 150)
1. Russia and the United States
2. Arctic Ocean
3. Pacific Ocean

Week 28 Day 3 (page 152)
1. A strait is a narrow waterway that connects two large bodies of water.
2. the Arctic Ocean and the Pacific Ocean
3. Answers may include that it is a small, remote island with few residents.

Week 28 Day 4 (page 153)
1. 2004
2. Answers should include the amount of ice is decreasing.
3. Answers should describe the amount in 2020 will be less than in 2016.

Week 29 Day 1 (page 155)
1. The United States
2. Ottawa
3. Winnipeg

Week 29 Day 3 (page 157)
1. France owned that land and its resources.
2. Quebec
3. Many died from disease. They lost their land and their way of life, and their language was changed.

Week 29 Day 4 (page 158)
1. four
2. Both the French and the British ruled Canada.

Week 30 Day 1 (page 160)
1. Maryland and Virginia
2. Atlantic Ocean
3. East coast; explanations may include that it is connected to the Atlantic Ocean, which is on the east coast.

Week 30 Day 3 (page 162)
1. An estuary is where ocean water meets freshwater.
2. Answers may include catch fish and crabs, look for oysters, and sail.
3. They can pick up trash and keep the rivers clean.

Week 30 Day 4 (page 163)
1. 18 million
2. 64,000 square miles (166,000 square km)
3. six states; answers may include that people and states have to work together.
4. Answers may include that if the watershed is dirty, the bay will also become dirty.

Week 31 Day 1 (page 165)
1. Arizona
2. Grand Canyon National Park
3. Little Colorado River, Kanab Creek, Havasu Creek, and Diamond Creek

Week 31 Day 3 (page 167)
1. It eroded the rocks.
2. It is one mile deep in some places.
3. It controls the flow of a river and helps get water to people.

Week 31 Day 4 (page 168)
1. The upper edge of the white band above the water should be traced.
2. There is less water; reasons may include that the water level is below the white band.

ANSWER KEY (cont.)

3. Answers may include that there will be less water or those states need to conserve water.

Week 31 Day 5 (page 169)

Picture should include people watering plants and using water for recreation.

Week 32 Day 1 (page 170)

1. New Mexico, Texas, Oklahoma, Colorado, Kansas, Nebraska, Wyoming, Montana, South Dakota, North Dakota
2. Saskatchewan, Manitoba, Alberta
3. North Dakota, South Dakota, Nebraska, Kansas

Week 32 Day 3 (page 172)

1. Settlers hunted too many bison.
2. Answers may include that American Indians lost their land, their food source, and their way of life.
3. It changed their way of life. They could not follow the bison anymore.

Week 32 Day 4 (page 173)

1. Answers may include that the tepee is cone-shaped, has a flap for an opening, and is held up by poles.
2. Answers may include that it could be taken down and moved quickly.
3. Answers may include that they are alike because they can be put up and taken down quickly, and they are different because tents today are not cone-shaped.

Week 33 Day 1 (page 175)

1. Omaha, Nebraska
2. Sacramento, California
3. Promontory, Utah

Week 33 Day 3 (page 177)

1. It made travel faster and safer.
2. Answers may include that more people could move there, and people who lived there could get mail and supplies faster.

Week 33 Day 4 (page 178)

1. train
2. ship
3. Answers may include that they would want to travel by train because it is fastest.

Week 34 Day 1 (page 180)

1. Midwest
2. South
3. Northeast
4. Three western states should be listed.

Week 34 Day 3 (page 182)

1. They might move for better weather, a new job, or to be near family.
2. the Northeast and the Midwest
3. Answers may include that it may need more schools, homes, roads, and resources.

Week 34 Day 4 (page 183)

1. Northeast: 1 million; Midwest: 1 million, South: 7 million; West: 4 million
2. The South and the West grew the most.
3. The Northeast and the Midwest grew the least.

Week 35 Day 1 (page 185)

1. Idaho, Wyoming, Montana
2. south
3. Grand Teton National Park

Week 35 Day 3 (page 187)

1. 1926
2. 1974
3. Answers may include that the ecosystem is more balanced or that the elk may be endangered next.

Week 35 Day 4 (page 188)

1. birds
2. amphibians
3. 61 more mammal species
4. Answers may include that there are many types of animals in the park.

Week 36 Day 1 (page 190)

1. Illinois
2. California
3. Idaho
4. Utah

Week 36 Day 3 (page 192)

1. Climate is what the weather is usually like.
2. It has a dry climate with long, sunny days and cool nights.
3. They both have mild climates where it does not get too cold.

Week 36 Day 4 (page 193)

1. Midwest
2. South and West
3. Midwest; 60°F
4. Answers should reference the chart.

Name: _____   Date: _____

MAP SKILLS RUBRIC
DAYS 1 AND 2

Directions: Evaluate students' activity sheets from the first two weeks of instruction. Every five weeks after that, complete this rubric for students' Days 1 and 2 activity sheets. Only one rubric is needed per student. Their work over the five weeks can be evaluated together. Evaluate their work in each category by writing a score in each row. Then, add up their scores, and write the total on the line. Students may earn up to 5 points in each row and up to 15 points total.

Skill	5	3	1	Score
Identifying Map Features	Identifies and uses all features on a map, including the title, legend, and compass rose.	Identifies and uses most features on maps.	Does not identify and use features on maps.	
Using Cardinal Directions	Uses cardinal direction to accurately locate places all or nearly all of the time.	Uses cardinal direction to accurately locate places most of the time.	Does not use cardinal directions to accurately locate places.	
Interpreting Maps	Accurately interprets maps to answer questions about it all or nearly all the time.	Accurately interprets maps to answer questions about it most of the time.	Does not accurately interprets maps to answer questions.	

Total Points: _____

Name: _____ Date: _____

APPLYING INFORMATION AND DATA RUBRIC
DAYS 3 AND 4

Directions: Complete this rubric every five weeks to evaluate students' Day 3 and Day 4 activity sheets. Only one rubric is needed per student. Their work over the five weeks can be evaluated together. Evaluate their work in each category by writing a score in each row. Then, add up their scores, and write the total on the line. Students may earn up to 5 points in each row and up to 15 points total. **Note:** Weeks 1 and 2 are map skills only and will not be evaluated here.

Skill	5	3	1	Score
Interpreting Texts	Correctly interprets texts to answer questions all or nearly all the time.	Correctly interprets texts to answer questions most of the time.	Does not correctly interpret texts to answer questions.	
Interpreting Data	Correctly interprets data to answer questions all or nearly all the time.	Correctly interprets data to answer questions most of the time.	Does not correctly interpret data to answer questions.	
Applying Information	Applies new information and data to known information about other locations or regions all or nearly all the time.	Applies new information and data to known information about other locations or regions most of the time.	Does not apply new information and data to known information about other locations or regions.	

Total Points: _____

Name: _____   Date:_____

MAKING CONNECTIONS RUBRIC
DAY 5

Directions: Complete this rubric every five weeks to evaluate students' Day 5 activity sheets. Only one rubric is needed per student. Their work over the five weeks can be evaluated together. Evaluate their work in each category by writing a score in each row. Then, add up their scores, and write the total on the line. Students may earn up to 5 points in each row and up to 15 points total. **Note:** Weeks 1 and 2 are map skills only and will not be evaluated here.

Skill	5	3	1	Score
Comparing One's Community	Makes meaningful comparisons of one's own home or community to others all or nearly all the time.	Makes meaningful comparisons of one's own home or community to others most of the time.	Does not make meaningful comparisons of one's own home or community to others.	
Comparing One's Life	Makes meaningful comparisons of one's daily life to those in other locations or regions all or nearly all the time.	Makes meaningful comparisons of one's daily life to those in other locations or regions most of the time.	Does not makes meaningful comparisons of one's daily life to those in other locations or regions.	
Making Connections	Uses information about other locations or regions to make meaningful connections about life there all or nearly all the time.	Uses information about other locations or regions to make meaningful connections about life there most of the time.	Does not use information about other locations or regions to make meaningful connections about life there.	

Total Points: _____

MAP SKILLS ANALYSIS

Directions: Record each student's rubric scores (page 202) in the appropriate columns. Add the totals, and record the sums in the Total Scores column. Record the average class score in the last row. You can view: (1) which students are not understanding map skills and (2) how students progress throughout the school year.

Student Name	Week 2	Week 7	Week 12	Week 17	Week 22	Week 27	Week 32	Week 36	Total Scores
Average Classroom Score									

APPLYING INFORMATION AND DATA ANALYSIS

Directions: Record each student's rubric scores (page 203) in the appropriate columns. Add the totals, and record the sums in the Total Scores column. Record the average class score in the last row. You can view: (1) which students are not understanding how to analyze information and data and (2) how students progress throughout the school year.

Student Name	Week 7	Week 12	Week 17	Week 22	Week 27	Week 32	Week 36	Total Scores
Average Classroom Score								

MAKING CONNECTIONS ANALYSIS

Directions: Record each student's rubric scores (page 204) in the appropriate columns. Add the totals, and record the sums in the Total Scores column. Record the average class score in the last row. You can view: (1) which students are not understanding how to make connections to geography and (2) how students progress throughout the school year.

Student Name	Week 7	Week 12	Week 17	Week 22	Week 27	Week 32	Week 36	Total Scores
Average Classroom Score								

DIGITAL RESOURCES

To access the digital resources, go to this website and enter the following code: 09457331. www.teachercreatedmaterials.com/administrators/download-files/

Rubrics

Resource	Filename
Map Skills Rubric	skillsrubric.pdf
Applying Information and Data Rubric	datarubric.pdf
Making Connections Rubric	connectrubric.pdf

Item Analysis Sheets

Resource	Filename
Map Skills Analysis	skillsanalysis.pdf skillsanalysis.docx skillsanalysis.xlsx
Applying Information and Data Analysis	dataanalysis.pdf dataanalysis.docx dataanalysis.xlsx
Making Connections Analysis	connectanalysis.pdf connectanalysis.docx connectanalysis.xlsx

Standards

Resource	Filename
Standards Charts	standards.pdf